U. Reischl C. Wittwer F. Cockerill (Eds.)

Rapid Cycle Real-Time PCR – Methods and Applications

Microbiology and Food Analysis

Springer

Berlin
Heidelberg
New York
Barcelona
Hong Kong
London
Milan
Paris
Tokyo

U. Reischl C. Wittwer F. Cockerill (Eds.)

Rapid Cycle Real-Time PCR – Methods and Applications

Microbiology and Food Analysis

With 76 Figures and 127 Tables

Dr. UDO REISCHL
Universitätsklinikum Regensburg
Institut für Medizinische Mikrobiologie und Hygiene
Franz-Josef-Strauß-Allee 11
93053 Regensburg, Germany

Prof. Dr. CARL WITTWER
University of Utah Medical School
Department of Pathology
Salt Lake City, UT 84132, USA

Prof. Dr. FRANKLIN COCKERILL
Division of Clinical Microbiology
Mayo Clinic
Rochester, Minnesota 55905, USA

ISBN 3-540-41881-4 Springer Verlag Berlin Heidelberg New York

Library of Congress applied for
Die Deutsche Bibliothek – CIP-Einheitsaufnahme
Rapid cycle real-time PCR : methods and applications ; microbiology and food analysis / Udo Reischl ... (ed.). - Berlin ; Heidelberg ; New York ; Barcelona ; Hong Kong ; London ; Mailand ; Paris ; Singapore ; Tokyo : Springer, 2001
ISBN 3-540-41881-4

Springer-Verlag Berlin Heidelberg New York
a member of BertelsmannSpringer Science+Business Media GmbH

http://www.springer.de/medizin

Printed in Germany

Cover Design: design & production, 69121 Heidelberg, Germany
Production: ProEdit GmbH, 69126 Heidelberg, Germany
Typesetting: TBS, 69207 Sandhausen, Germany
Printed on acid free paper SPIN 10832920 18/3130 Re – 5 4 3 2 1 0

Table of Contents

Introduction for Microbiology Volume of Rapid Cycle Real-Time PCR 1
Franklin Cockerill

Applications and Challenges of Real-Time PCR for the Clinical Microbiology Laboratory 3
Franklin R. Cockerill, III*, James R. Uhl

Part I
Bacteriology

Rapid Detection and Simultaneous Differentiation of *Bordetella pertussis* and *Bordetella parapertussis* in Clinical Specimens by LightCycler PCR 31
Udo Reischl, Katrin Kösters, Birgit Leppmeier, Hans-Jörg Linde, Norbert Lehn

Rapid Detection and Simultaneous Differentiation of *Legionella* spp. and *L. pneumophila* in Potable Water Samples and Respiratory Specimens by LightCycler PCR 45
Nele Wellinghausen, Olfert Landt, Udo Reischl

Homogeneous Assays for the Rapid PCR detection of *Burkholderia pseudomallei* 16s rRNA gene on a Real-Time Fluorometric Capillary Thermocycler 59
Eric P-H Yap, Soon-Meng Ang, Shirley G-K Seah, Seng-Meng Phang

Rapid Detection of Toxigenic *Corynebacterium diphtheriae* by LightCycler PCR 71
Udo Reischl, Elena Samoilovich, Valentina Kolodkina, Norbert Lehn, Hans-Jörg Linde

Rapid Detection of Resistance Associated Mutations in *Mycobacterium tuberculosis* by LightCycler PCR 83
Maria J. Torres, Antonio Criado, Jose C. Palomares, Javier Aznar

Duplex LightCycler PCR Assay for the Rapid Detection of Methicillin-Resistant *Staphylococcus aureus* and Simultaneous Species Confirmation 93
UDO REISCHL, HANS-JÖRG LINDE, BIRGIT LEPPMEIER, NORBERT LEHN

Rapid Detection of Group B Streptococci Using the LightCycler Instrument 107
DANBING KE, CHRISTIAN MÉNARD, FRANÇOIS J. PICARD, MICHEL G. BERGERON

Rapid Detection and Quantification of *Chlamydia trachomatis* in Clinical Specimens by LightCycler PCR 115
HEIDI WOOD, UDO REISCHL, ROSANNA W. PEELING

Quantification of *Toxoplasma gondii* in Amniotic Fluid by Rapid Cycle Real-Time PCR 133
FRANÇOIS DELHOMMEAU, FRANÇOIS FORESTIER

Genospecies-Specific Melting Temperature of the *recA* PCR Product for the Detection of *Borrelia burgdorferi* sensu lato and Differentiation of *Borrelia garinii* from *Borrelia afzelii* and *Borrelia burgdorferi* sensu stricto ... 139
JOHANNA MÄKINEN, QIUSHUI HE, MATTI K. VILJANEN

Rapid and Specific Detection of *Coxiella burnetii* by LightCycler PCR 149
MARKUS STEMMLER, HERMANN MEYER

Molecular Identification of *Campylobacter coli* and *Campylobacter jejuni* Using Automated DNA Extraction and Real-Time PCR with Species-Specific TaqMan Probes 155
LYNN A. COOPER, LUKE T. DAUM, ROBERT ROUDABUSH, KENTON L. LOHMAN

Rapid and Specific Detection of *Plesiomonas shigelloides* Directly from Stool by LightCycler PCR 161
JIN PHANG LOH, ERIC PENG HUAT YAP

Quantification of Thermostable Direct Hemolysin-Producing *Vibrio parahaemolyticus* from Foods Assumed to Cause Food Poisoning Using the LightCycler Instrument 171
YOSHITO IWADE, AKINORI YAMAUCHI, AKIRA SUGIYAMA, OSAMU NAKAYAMA, NORICHIKA H. KUMAZAWA

Part II
Detection of Fungal Pathogens

Quantification and Speciation of Fungal DNA in Clinical Specimens Using the LightCycler Instrument 179
Juergen Loeffler, Holger Hebart, Norbert Henke, Kathrin Schmidt, Hermann Einsele

Part III
Virology

Development, Implementation, and Optimization of LightCycler PCR Assays for Detection of Herpes Simplex Virus and Varicella-Zoster Virus from Clinical Specimens 189
Thomas F. Smith, Mark J. Espy, Arlo D. Wold

Qualitative Detection of Herpes Simplex Virus DNA in the Routine Diagnostic Laboratory 201
Harald H. Kessler, Gerhard Mühlbauer, Evelyn Stelzl, Egon Marth

Use of Real-Time PCR to Monitor Human Herpesvirus 6 (HHV-6) Reactivation . . . 211
Junko H. Ohyashiki, Kazuma Ohyashiki, Kohtaro Yamamoto

Simultaneous Identification of Five HCV Genotypes in a Single Reaction 217
Olfert Landt, Ulrich Lass, Heinz-Hubert Feucht

Rapid Detection of West Nile Virus 227
Olfert Landt, Jasmin Dehnhardt, Andreas Nitsche, Gary Milburn, Shawn D. Carver

Use of Rapid Cycle Real-Time PCR for the Detection of Rabies Virus 235
Lorraine McElhinney, Jason Sawyer, Christopher J. Finnegan, Jemma Smith, Anthony R. Fooks

Development of One-Step RT-PCR for the Detection of an RNA Virus, Dengue, on the Capillary Real-Time Thermocycler 241
Boon-Huan Tan, E'ein See, Elizabeth Lim, Eric Peng-Huat Yap

Part IV
Detection of Genetically Modified Organisms (GMO)

Variation Analysis of Seven LightCycler Based Real-Time PCR Systems to Detect Genetically Modified Products (RRS, Bt176, Bt11, Mon810, T25, Lectin, Invertase) 251
ISABELLE DAHINDEN, ANDREAS ZIMMERMANN, MARIANNE LINIGER, URS PAULI

Abbreviations

Bp	Base pair
EDTA	Ethylenediamine Tetraacetic Acid
F	Fluorescein
LCRed640	LightCycler Red 640 Fluorescent Dye
LCRed705	LightCycler Red 705 Fluorescent Dye
Nt	Nucleotide(s)
P	Phosphate
PBS	Phosphate Buffered Saline
SDS	Sodium Dodecylsulfate
TE	1xTE Buffer Contains 0.01 M Tris, 0.001 M EDTA, pH 8.0
TE'	1xTE' Buffer Contains 0.01 M Tris, 0.0001 M EDTA, pH 8.0
TEN (=STE)	1xTEN Buffer Contains 0.01 M Tris (pH 8.0), 150 mM NaCI, 0.05% Tween 20
T_m(°C)	Melting Temperature
n-n	nearest neighbor
FRET	Fluorescence Resonance Energy Transfer
ΔH	Enthalpy
ΔS	Entropy
ΔG	Free Energy
TPMT	Thiopurine Methyltransferase
XO	Xanthine Oxidase
AA	Amino Acid
ORF	Open Reading Frame

Introduction for Microbiology Volume of Rapid Cycle Real-Time PCR

Several commercial testing platforms have become available or soon will be available which combine rapid cycle polymerase chain reaction with fluorescent probe detection of amplified target nucleic acid in the same closed vessel. This technology is referred to as rapid cycle real-time PCR.

One testing platform which exploits this technology, the LightCycler instrument (Roche Diagnostics Corporation), has become particularly popular. The LightCycler system was the first system available that permitted the operator to evaluate target amplification after each PCR cycle. Recently an automated nucleic acid extraction system, the MagNA Pure LC (Roche Diagnostics Corporation), has been developed to complement the LightCycler system. The combination of the MagNA Pure LC and LightCycler system now permits complete automation of extraction, purification, concentration, amplification and detection of target nucleic acid from patient samples.

Although the LightCycler system was initially intended for research purposes, it became quickly obvious to many microbiologists, that the system had immense potential for diagnostic microbiology testing. The LightCycler-MagNA Pure LC system could not only replace cumbersome and work intense procedures required for conventional, "home-brewed" PCR assays, but as well could replace standard antigen or culture-based assays. As an example, the accepted diagnostic approach for detecting Group A Streptococcus (*Streptococcus pyogenes*) from throat swabs is to screen for group A streptococcal antigens using a commercial direct immunoassay. If the result for this "rapid antigen" assay is negative, a culture is performed. This two-step procedure follows guidelines provided by the Infectious Diseases Society of America and other professional organizations and is required because current rapid antigen assays lack sensitivity. We have developed a LightCycler assay which is ~10% more sensitive and approximately half the cost ($1-$2 vs $2-$3) and requires much less time to perform (<1 hr vs up to 48 h) than the combined antigen-culture assay. In addition to accuracy, speed and cost issues, the LightCycler system and other rapid cycle real-time PCR platforms have other advantages over conventional PCR methods. These include decreased risk for contamination (specimen or amplicon) and adaptability for a wide range of testing. The containment of the PCR reaction and amplicon detection in the same closed vessel, as is possible with the LightCycler system, lessens the possibility of amplicon contamination including amplicon "carryover" from previous assay runs. As well, the possibility of specimen to specimen contamination is essentially eliminated. The versatility for testing is exemplified by applications which identify or quantitate organisms, identify virulence genes, or determine mutations associated with antimicrobial resistance.

All of these remarkable performance characteristics have been demonstrated for LightCycler assays for many other agents of infectious disease besides Group A Streptococcus. Many of these assays are described in detail in the chapters that follow this introduction and are intended for the detection or quantification of human pathogens from human samples or foodstuffs. In some cases, specimen processing, including nucleic acid extraction, purification and concentration, using the automated MagNA Pure LightCycler system, is also described. I would like to take this opportunity to thank all of the authors who have developed the chapter for this book. These professionals are recognized international experts in clinical microbiology who have developed, verified and validated these LightCycler assays.

Rapid cycle real-time PCR is a revolutionary testing platform for the contemporary microbiology laboratory. It is likely that this technology will significantly change the way we identify or quantify pathogens and determine their virulence factors or antimicrobial resistance. I am certain you will find the information provided herein very interesting and useful and also quite exciting.

October 2001 FRANKLIN COCKERILL for the Editors

Applications and Challenges of Real-Time PCR for the Clinical Microbiology Laboratory

Franklin R. Cockerill, III*, James R. Uhl

Introduction

Recently, significant improvements have occurred in molecular diagnostic testing methods, especially those that incorporate the polymerase chain reaction (PCR). Several commercial testing platforms have become available, or soon will be available, which combine PCR amplification and probe detection of target nucleic acid in the same closed reaction vessel. This permits rapid, dynamic assessment of PCR products (amplicons) and lessens the possibility of contamination by extraneous nucleic acid.

These testing platforms have been referred to as "real-time" PCR, implying that PCR amplicons can be detected in real time. In actuality, "real-time" refers to the detection of amplicons after each PCR cycle. Other descriptions of this technology include "kinetic PCR" and "homogeneous PCR."

The objectives of this paper are threefold. First, a brief review of conventional PCR and real-time PCR testing formats will be provided. Following this, advantages of real-time PCR formats will be presented. Finally, the potential applications and challenges of real-time PCR testing methods in the clinical microbiology laboratory will be discussed.

Conventional PCR-Based Testing Formats Used in the Clinical Microbiology Laboratory

Conventional or "home brew" (in-house developed) PCR testing formats generally involve at least three and sometimes four steps. In the first step, "target" nucleic acid is extracted from the patient's specimen. Frequently, the "target" nucleic acid is unique to a microorganism and therefore permits identification or quantification of that organism. Alternatively, nucleic acid may be targeted, which is part of a gene that encodes for antimicrobial resistance or is a segment of nucleic in which mutations occur that are associated with antimicrobial resistance.

* Franklin R. Cockerill, III (✉) (e-mail: cockerill.franklin@mayo.edu)
Division of Clinical Microbiology, Mayo Clinic, Hilton 470B, 200 First Street S.W., Rochester, MN 55905, USA

In the second step, the polymerase chain reaction (PCR) is carried out. Conventional thermocyclers require 1–2 h to complete 30–40 "thermal" cycles. This number of thermal cycles is frequently required for the detection of nucleic acid from microorganisms in human samples. With each thermal cycle, the temperature is raised to ~90°C to denature double-stranded DNA (denaturation step), lowered to ~55°C to allow hybridization of PCR primers to single-stranded DNA (annealing step), and raised to ~70°C at which the thermal-stabile Taq polymerase enzyme can function and produce a complementary strand of DNA (elongation step).

For the third step of conventional PCR testing, the amplicon is "sized" by electrophoresing it through an agarose gel and staining with ethidium bromide. The distance the amplicon migrates through the gel is compared to the distance molecular weight standards migrate through the gel. A molecular weight is assigned (amplicon is sized) based on this comparison.

A fourth step is usually performed to confirm the amplicon as the correct amplicon. Many testing formats now eliminate the third step and only perform this fourth confirmatory step. The confirmation of amplicon was traditionally accomplished by manual-intense solid phase probe-hybridization formats such as Southern blot analysis; newer formats such as enzyme-linked hybridization capture assays have simplified this process.

In summary, these four steps for conventional PCR testing frequently required at least 2 days and sometimes 3 days in our laboratory to complete. Because of the hands-on technical expertise required of these steps, PCR testing has been limited to highly specialized clinical microbiology laboratories.

Newer PCR-Based Testing Formats

Table 1 displays both conventional and newer PCR-based testing formats. Included in this table is information about the principle of the technique, time required to complete the procedure (analytic turnaround time or analytical TAT), the approximate cost per test, commercial availability of kits or analyte-specific reagents (ASRs), and potential for contamination by either extraneous unamplified nucleic acid or amplified nucleic acid (amplicons). Contamination of specimens by extraneous nucleic acid occurs from one specimen to another specimen (cross-contamination) and often is the result of pipetting errors or splashes. Contamination of specimens by amplified nucleic acid (amplicons) is frequently of much greater concern and occurs when amplicons generated from a previous PCR assay contaminate specimen processing areas (amplicon carryover).

As indicated in the previous section, an important step in improving PCR-based testing platforms was the introduction of more rapid, user-friendly methods for amplicon detection, especially nonradioactive colorimetric enzyme-linked assays. One commercial adaptation of enzyme-linked amplicon detection assays, the AMPLICOR System (Roche Diagnostics Corporation, Indianapolis, IN), incorporates microwell plates in a semi-automated format. Another important feature of the AMPLICOR system is that all amplicons are "sterilized," that is, all amplicons are chemically altered so that further amplification cannot occur. The AMPLICOR assay incorporates deoxyuridine triphosphate (dUTP) in place of deoxythymidine triphosphate (dTTP) into amplicons. Uracil-*N*-glycosylase (UNG) catalyzes the cleavage of deoxyuridine containing DNA at deoxyuridine residues by opening the deoxyribose chain at the C1 position. UNG is added to all PCR master mixes and therefore will disrupt ("sterilize") any contaminating (carryover) amplicons. UNG is inactivated at temperatures above 55°C and subsequently target amplicons formed with the PCR reaction will not be destroyed. The

Table 1. Summary of PCR-based methods reported for detection or quantitation of human pathogens or for assessment of antimicrobial resistance in human pathogens[a]

Name of method	Principle	TAT[b]	Cost[c]	Commercial kit or ASRs available	Potential for contamination[g]
PCR gel electrophoresis	Amplicon sized in gel	6–8 h	~$5	No	Moderate to high[h]
PCR probe hybridization	Amplicon is detected by hybridization probe by Southern blotting, dot blots, or liquid hybridization formats	8–24 h	~$5–$10	No	Moderate to high[h]
PCR AMPLICOR-COBAS	Amplicon is biotin-labeled and detected by enzyme-linked assay (AMPLICOR); PCR and detection can be combined in an automated method (COBAS)	4 h	~$35 (qualitative analysis); ~$55 (quantitative analysis)	Qualitative tests for hepatitis C, *M. tuberculosis*, *C. trachomatis*, *N. gonorrhoeae*; quantitative tests for CMV, hepatitis B, hepatitis C, HIV-1 (Roche Diagnostics Corporation)	Low (amplicons are sterilized)
PCR restriction fragment polymorphism (RFLP)	Amplicon fragmented by restriction endonucleases; restriction patterns useful for identifying point mutations associated with antimicrobial resistance or for differentiating resistance genes with similar DNA sequences	8–10 h	~$5	No	Moderate to high[h]
PCR single-strand conformation polymorphism (SSCP)	Amplicons are denatured and electrophoresed. Mobility shifts in high-resolution nondenaturing polyacrylamide gels are discernible for single-stranded mutated DNA vs single-stranded wild-type DNA	2 days	~$10	No	Moderate to high[h]
PCR DNA sequencing	The nucleic acid sequence of amplicons is determined	1–2 days	~$10	For HIV mutation screening: TruGene HIV-1 assay (Visible Genetics, Inc., Toronto, Ontario, Canada)	Low to moderate

Table 1. *Continued*

Name of method	Principle	TAT[b]	Cost[c]	Commercial kit or ASRs available	Potential for contamination[g]
PCR heteroduplex (HDP)	DNA from a universal heteroduplex generator and the test strain are denatured simultaneously in the same reaction mixture and then allowed to reanneal. As an example, the heteroduplex generator is constructed to have a 4-bp deletion. When separate strands of DNA from the heteroduplex generator and the test strain reanneal, four separate hybridizations can occur. Two of these hybridizations (heteroduplexes) will result in double-stranded DNA with a "bubble" occurring where no base-pair matches can hybridize. These hybrids will migrate as a single band when electrophoresed in a high-resolution gel. Two other bands will occur due to hybridization reactions (homoduplexes) which result in the formation of the double-stranded DNA of the heteroduplex generator and the test strain. If mutations are present in the test strain, the migratory positions of the homoduplexes may vary compared to those of homoduplexes formed when no mutations are present	~8–10 h	~$5	No	Moderate to high[h]
PCR cleavase-fragment polymorphism (CFLP)	Labeled fragments of DNA are denatured. Single strands of DNA assume folded hairpin-like structures; subtle differences in the sequence of fragments can form different folds. The Cleavase I enzyme cleaves at the 5′ side of these structures, at the junction between duplexed and single-stranded regions. Separation and detection of the resulting fragments create signature banding patterns that can be compared to detect differences between mutated and wild-type DNA sequences	~1 day	NA	No[d]	Moderate to high[h]
PCR line probe assay (LiPA)	Biotinylated PCR amplicons are hybridized to immobilized probes on a nitrocellulose strip. The biotin group in the hybridization complex is then revealed by incubation with a streptavidin-alkaline phosphatase complex and chromogenic compounds	~8 h	~$80	For HCV genotyping and mutation detection in *rpoB* gene of *M. tuberculosis* associated with rifampin resistance (Innogenetics)	Low to moderate
PCR non-isotopic RNase cleavage assay (NIRCA)	RNA/RNA duplexes are produced from target (mutated) DNA and wildtype DNA. Unpaired bases in RNA/RNA duplexes are selectively cleaved by RNase treatment. Cleavage products are analyzed by gel electrophoresis	~1 day	~$50	No[e]	Moderate to high[h]

Table 1. *Continued*

Name of method	Principle	TAT[b]	Cost[c]	Commercial kit or ASRs available	Potential for contamination[g]
Rapid homogeneous PCR probe detection (real-time PCR)	Rapid thermocycling combined with homogeneous chemistry (PCR and amplicon detection are performed in the same reaction vessel) permits real-time detection of target nucleic acid in a closed system	30 min to 1–2 h	NA	No[f]	Low

[a] ASRS, analyte specific reagents; NA, not available, PCR, polymerase chain reaction.
[b] TAT, turnaround time for analytic procedure; TAT is increased for RNA viruses requiring an initial reverse transcriptase step. Specimen processing time is not included in TAT determination.
[c] Cost in US dollars. Only cost of reagents included per test sample; initial capital outlay for a conventional thermocycler (~\$5,000–\$10,000) or rapid thermocyclers (for real-time PCR assays) (\$25,000–\$50,000), electrophoresis equipment (~\$3,000) or in some instances a DNA sequencer (~\$60–\$150,000, depending on capacity and method) or COBAS instrument (~\$60,000) may be required, depending on the assay. RNA viruses may require a reverse transcriptase step which adds to the expense of the assay.
[d] ASRs may be available in the future for infectious diseases testing from Third Wave Technologies, Inc.
[e] ASRs may be available in the future for infectious diseases testing from Ambion, Inc., Austin, TX.
[f] ASRs may be available in the future for infectious diseases testing by LightCycler (Roche Molecular Biochemicals, Indianapolis, IN), SmartCycler (ARIDIA, Quebec, Canada), and MX4000 (Stretagene, Austin, TX) and others.
[g] Contamination of samples by extraneous unamplified nucleic acid can occur with any system during initial specimen processing and is usually specimen-to-specimen (cross-contamination of patients' specimens). Contamination of specimens by amplified nucleic acid (amplicons) is frequently of greater concern. The potential for amplicon contamination should be significantly reduced for all methods if amplicons are "sterilized", e.g., incorporation of deoxyuridine (dUTP) in amplicons and subsequent cleavage of DNA by uracil-N-glycosylase (UNG).
[h] Low if PCR performed only on high copy numbers (bacterial colonies or large quantities of infectious agents in specimens, e.g., high viral loads).

AMPLICOR system can also be used with the COBAS instrument. The COBAS instrument automates both the PCR and detection steps and has been used for qualitative and quantitative analyses of HIV and HCV.

As shown in Table 1, specialized gel detection formats were also developed to identify mutations in gene fragments. In clinical microbiology, most of these methods have been applied in home brew assays for mutation detection in genes associated with antimicrobial resistance, especially antimycobacterial resistance, and include restriction fragment length polymorphism (RFLP), single-strand conformation polymorphism (SSCP), and heteroduplex analyses. While all of the mutation screening methods are relatively inexpensive, they are manual, work intensive, time-consuming, and have the potential for contamination.

Several commercial methods, which are designed to be used in conjunction with PCR methods, have recently been developed to detect microorganisms or mutations within their genomes and are also shown in Table 1. These include cleavase-fragment polymorphism (Third Wave Technologies, Madison, WI), line probe (Innogenetics, Ghent, Belgium), and the RNase cleavase (Ambion, Inc., Austin, TX) assays.

Non-PCR-Based Testing Formats

Target nucleic acid amplification techniques other than PCR that are coupled with probe detection formats have been commercially developed. These methods will not be discussed in detail and include strand displacement amplification, a method involving amplification of target DNA transformed with nickable *Hinc*II sites (SDA, Becton Dickinson and Company, Rutherford, NJ), and two transcription-based amplification assays, nucleic acid sequenced-based amplification (NASBA, Organon Teknika Corporation, Durham, NC), and transcription-mediated amplification (GenProbe, Inc., San Diego, CA).

Still other molecular amplification methods are based on either "probe" or "signal" amplification technology. In the Ligase Chain Reaction (Abbott Laboratories, Abbott Park, IL), both target DNA and probes hybridized to target nucleic acid are "amplified." For this assay, adjacent oligonucleotide probes are annealed to target DNA and subsequently ligated; these ligated probes and the target DNA serve as substrates for subsequent ligation reactions. Examples of probe only or signal-only amplification formats are shown in Table 2. Probe amplification platforms include cycling probe technology (ID Biochemical, Vancouver, BC, Canada), the hybrid capture system (Digene, Gaithersberg, MD), and the QB replicase technology (Vysis, Inc., Naperville, IL). The branched signal amplification method (b-DNA, Bayer Corporation, Tarrytown, NY) is a signal amplification system. Another recently developed signal amplification method is linear signal amplification (Third Wave Technologies).

Except for the Ligase Chain Reaction, probe and signal amplification methods do not amplify target nucleic acid and therefore contamination by amplified target DNA (amplicons) is not an issue. However, small amounts of contaminating target DNA during sample preparation (cross-contamination of patient specimens) could theoretically result in false-positive results and amplified signal or probe could contaminate adjacent samples in open tube systems. This aside, unlike PCR amplicons that are not sterilized, amplified signal or probe cannot be further amplified in subsequent analyses.

Table 2. Summary of probe-based or signal-based amplification methods reported for detection or quantitation of human pathogens or for assessment of antimicrobial resistance in human pathogens[a]

Name of method	Principle	TAT[b]	Cost[c]	Commercial kit or ASRs available	Potential for contamination[e]
b-DNA	Target DNA is hybridized to capture probes. Additional probes (target probes) hybridize with target DNA. A branched DNA (bDNA) molecule hybridizes with the target probe. Finally, enzyme-labeled probes are hybridized to the bDNA molecules. The end result is signal amplification	6–8 h	~$100	Quantitative tests for HIV and HCV (Bayer Diagnostics, Berkeley, CA)	Low

Table 2. *Continued.*

Name of method	Principle	TAT[b]	Cost[c]	Commercial kit or ASRs available	Potential for contamination[e]
Linear signal amplification (LSA)	Two overlapping oligonucleotides hybridize to target DNA. The downstream oligonucleotide is a signal probe; the 5′ end of this probe is noncomplementary to the target DNA and forms an unpaired "flap." Cleavase VIII removes the flap, permitting a fluorescent compound on the 5′ end of the flap to be excited. Fluorescent-labeled cleavage products are separated in a denaturing polyacrylamide gel and detected by scanning. The reaction is cycled (additional flaps are hybridized with target DNA and cleaved) at the same temperature	NA	NA	No[d]	Low
Cycling probe technology (CPT)	A probe is hybridized to the target nucleic acid. An enzyme cleaves the attached probe without cleaving the target sequence. Probe fragments are detected on a membrane. The reaction is cycled at the same temperature; fragmented probe accumulates and target DNA serves as template for subsequent probe hybridization	90 min	~$10	For *mecA* gene detection: Velogene MRSA (ID Biomedical)	Low
Hybrid capture system	Target DNA hybridizes with a specific RNA probe. The resultant RNA-DNA hybrids are captured onto a tube coated with antibodies specific for RNA-hybrids. Immobilized hybrids are then reacted with enzyme-conjugated antibodies, which are then detected with a chemiluminescent substrate. Because multiple antibodies can bind to single-target sequence, "signal amplification" occurs	4 h	~$45	Available for human papilloma virus detection and typing (Digene)	Low
Qβ replicase	Poly (dG) and poly (dA) probes hybridize with target nucleic acid. A target-specific probe substrate for Qβ replicase, MDV-1, is also hybridized with the target nucleic acid. The poly (dG) and poly (dA) probes bind to paramagnetic beads which "capture" the target nucleic acid. The captured nucleic acid is exposed to Qβ replicase, resulting in amplification of the target specific Qβ replicase substrate (probe)	NA	NA	No	Low

[a] ASRS, analyte specific reagents; NA, not available; PCR, polymerase chain reaction; b-DNA to branched chain DNA.
[b] TAT, turnaround time for analytic procedure; TAT is increased for RNA viruses requiring an initial reverse transcriptase step. Specimen processing time is not included in TAT determination.
[c] Cost in US dollars. Only cost of reagents included per test sample; initial capital outlay for equipment for bDNA assay (Bayer System 340 bDNa Analyzer, (~$80,000)) not included; RNA viruses may require a reverse transcriptase step which adds to the expense of the assay.
[d] ASRs may be available in the future for infectious diseases testing from Third Wave Technologies, Inc.
[e] Contamination of specimen by extraneous nucleic acid can occur with any system during initial specimen processing and is usually specimen-to-specimen (cross-contamination of patient specimen).

Real-Time PCR Testing Formats Used in the Clinical Microbiology Laboratory

Real-time PCR combines amplification of target DNA with detection of amplicons in the same closed vessel. By virtue of the closed system, the risk of carry-over or contamination by unsterilized amplicons is significantly reduced. Real-time PCR formats have recently been coupled with rapid thermocycling in a single system. In addition to detection of target nucleic acid, real-time PCR instruments can quantitate target nucleic acid and differentiate alleles (determine sequence variation or point mutation). As well, by combining two or more different PCR reactions (multiplexing), multiple targets can be assessed in the same sample in the same reaction vessel. For allelic variation, probe:target duplexes in the patient's specimen are compared to wild-type probe:target duplexes by melting curve analysis. Probe:target duplexes will melt at different temperatures for wild-type target vs mutated (allelic) target.

Commercial real-time PCR instruments that have recently become available or should be available soon, according to the manufacturer, include the ABI PRISM 5700 and 7700 instruments (Applied Biosystems, Foster City, CA), iCyler (BioRad, Hercules, CA), LightCycler instrument (Roche Molecular Biochemicals, Indianapolis, IN), SmartCycler (Cepheid, Sunnyvale, CA), MX4000 (Strategene, La Jolla, CA), and Roto-Gene (Corbett Research, Sydney, Australia). The major characteristics of these instruments are summarized in Table 3.

Table 3. Instrumentation available for real-time PCR

Name of instrument	Manufacturer	Automated extraction	Thermal cycling format	Detection format[a]	TAT[b]	Data observation	Reaction capacity
GeneAmp 5700 and Prism 7700	Applied Biosystems	Yes (ABI Prism 6700)	Conventional[c]	TaqMan	~2 h	End-point	96/run
iCycler iQ	BioRad	No	Conventional[c]	TaqMan, FRET, or Molecular Beacons	~2 h	Each PCR cycle	96–384/run
LightCycler	Roche Diagnostics	Yes (MagNA Pure LC)	Ambient air cooling	TaqMan, FRET, or Molecular Beacons	~20 min to 1 h	Each PCR cycle	32/run
SmartCycler	Cepheid	No	Ceramic heating plate	TaqMan, FRET, or Molecular Beacons	~40 min to 1 h	Each PCR cycle	16–96/run
MX4000	Stratagene	No	Conventional[c]	Molecular Beacons	~90 min	Each PCR cycle	96/run
Rotor Gene	Corbett Research	No	Ambient air cooling	TaqMan, FRET, or Molecular Beacons	~50 min	Each PCR cycle	32/run

[a] Theoretically, TaqMan, FRET, or Molecular Beacon fluorophore probes could be used with each system; however, some manufacturers prefer certain fluorophore probes over others. In these cases, the preferred fluorophore probes are listed.
[b] TAT, analytical turnaround time; does not include specimen preparation, i.e., lysis of organisms, isolation, and purification of nucleic acid.
[c] Conventional refers to heating block thermal cycler.
[d] FRET, fluorescent resonance energy transfer.

Probe detection formats, which have been most frequently adapted to real-time instruments, include TaqMan (Roche Diagnostics Corporation), FRET (Fluorescent Resonance Energy Transfer, Roche Molecular Biochemicals), and Molecular Beacons (vendors available through www.molecularbeacons.com). The principles of these techniques are shown in Figs. 1–3.

Rapid thermocycling in real-time PCR instruments is possible by virtue of (a) rapid air exchange or rapid thermal conductivity through solid phase material surrounding the reaction vessel and (b) high surface-to-volume ratio of the PCR mix. The latter property is facilitated by narrow elongated reaction cuvettes.

The ABI GeneAmp 5700 instrument was the first real-time PCR instrument that was commercially available. This high throughput system, along with the newer version, the PRISM 7700, use a heating block for thermocycling and in our experience requires approximately 2 h to complete. Furthermore, although PCR and probe hybridization occur in real-time with these instruments, the hybridization data is not available to the user until all cycles of PCR are completed. In contrast, for all of the other systems shown in Table 3, real-time observation of hybridization data is possible with each PCR-probe hybridization cycle.

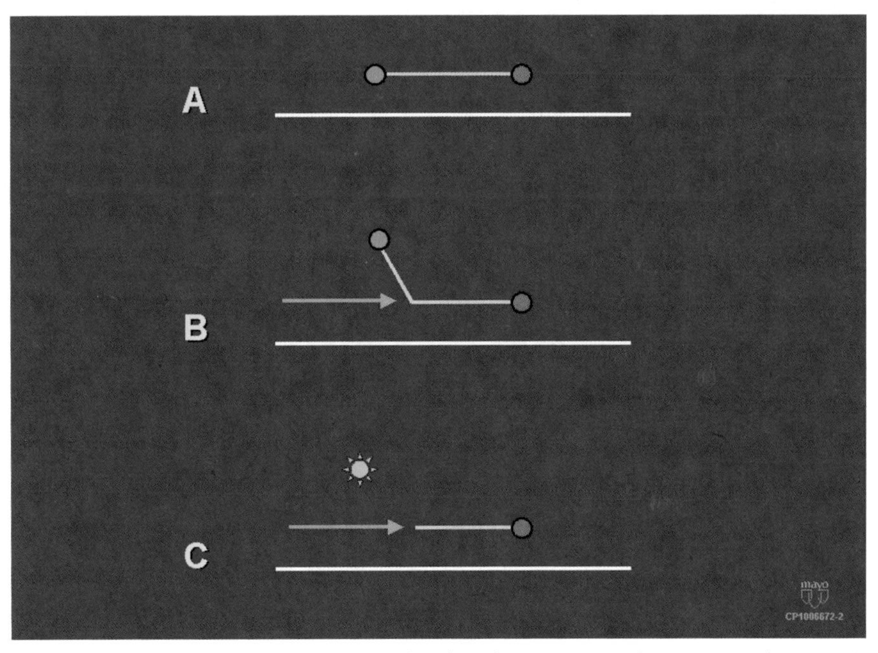

Fig. 1A–C. Principles of TaqMan probes. The 5′ nuclease activity of Taq DNA polymerase is exploited to cleave a TaqMan probe during PCR. **A** The TaqMan probe is shown annealed to the target DNA. The probe contains a reporter dye at the 5′ end of the probe (*dark green circle*) and a quencher dye at the 3′ end of the probe (*red circle*). **B** During the PCR reaction, a complementary strand of DNA is synthesized and the 5′ exonuclease activity of Taq DNA polymerase excises the reporter dye. **C** Fluorescence of the reporter dye (indicated by *asterisk* and *bright green color* of reporter dye) occurs as a result of separation of the reporter dye from the quencher dye

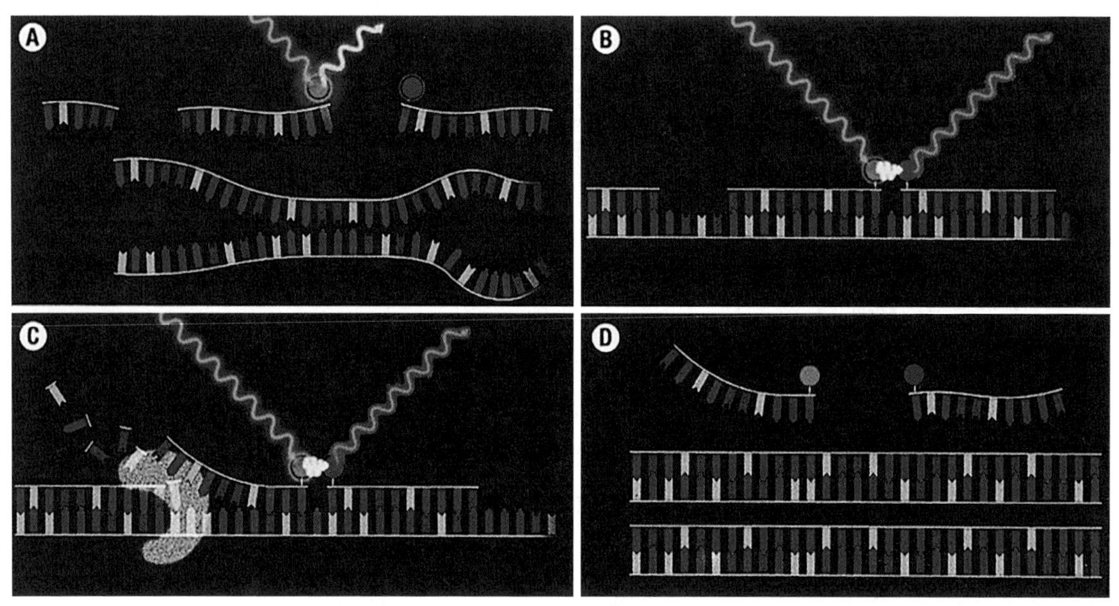

Fig. 2. Principles of FRET (fluorescent resonance energy transfer) using the LightCycler instrument. *Panel A* shows the essential components using fluorescence-labeled oligonucleotides as FRET probes: two different oligonucleotides (labeled) and the amplification product. Probe 1 bears a fluorescein label at its 3′ end while probe 2 is labeled with another label (LightCycler Red 640) at its 5′ end. The sequences of the two probes are selected so that they can hybridize to the amplified DNA fragment in a head-to-tail arrangement, thereby bringing the two fluorescent dyes into close proximity (*panel B*). The first dye (fluorescein) is excited by the Light Cycler's light source and emits green fluorescent light at a slightly longer wavelength. When the two dyes are in close proximity, the energy thus emitted excites the LightCycler Red 640 attached to the second probe that subsequently emits red fluorescent light at an even longer wavelength. This energy transfer, referred to as FRET, is highly dependent on the spacing between the two dye molecules. Only if the molecules are in close proximity (between 1 and 5 nucleotides) is the energy transferred efficiently. The intensity of the light emitted by the LCRed640 is measured in Channel 2 (640 nm) of the LightCycler's optics. The increasing amount of measured fluorescence is proportional to the increasing amount of DNA generated during the ongoing PCR process. Since LCRed640 only emits a signal when both oligos are hybridized, fluorescence is measured after the annealing step (*panel B*). Hybridization does not take place during the denaturation phase of the PCR (*panel A*), and fluorescence cannot be detected at 640 nm. After annealing, the temperature is raised and the hybridization probe is displaced by the Taq DNA polymerase (*panel C*). At the end of the elongation step, the PCR product is double-stranded and the probes are too far apart to allow FRET (*panel D*) (used with permission from Roche Molecular Biochemicals)

The LightCycler system (Roche Molecular Biochemicals) was the first system to permit the operator real-time data analyses and to be applied for diagnosis of infectious agents. The LightCycler system uses rapid air exchange as compared with the heating block used in the ABI GeneAmp 5700 and PRISM 7700 and analyzes smaller numbers of samples than either of these systems (32 vs 96). This latter property and a shorter thermocycling time permit testing in a more continuous flow as com-

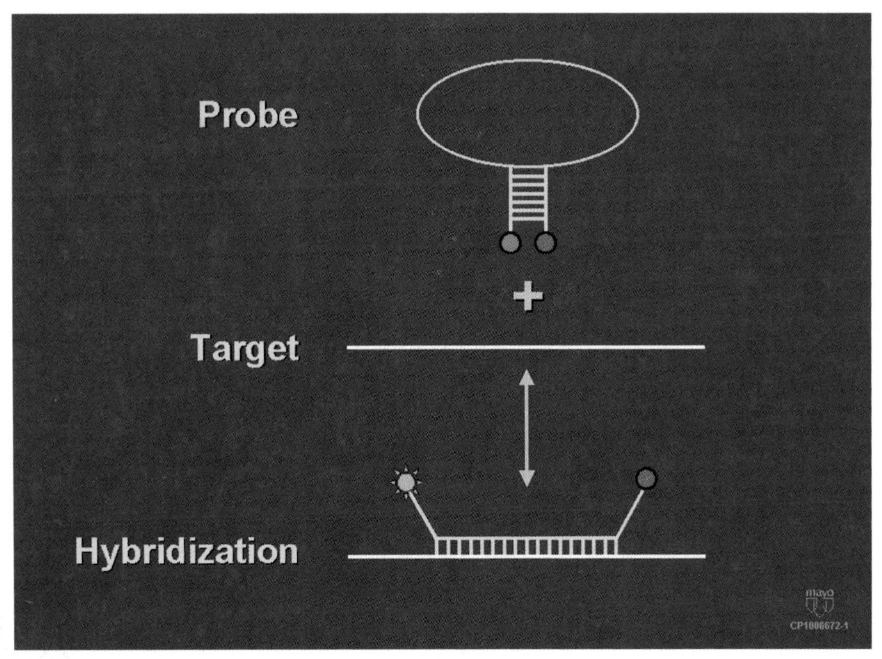

Fig. 3. Principle of molecular beacon probes. The molecular beacon probe in its hairpin form shown at the top of the figure is nonfluorescent because the stem hybrid keeps the fluorophore (reporter dye) close to the quencher dye. When the probe sequence in the loop hybridizes to its target as shown at the bottom of the figure, a rigid double helix is formed. This conformational reorganization separates the quencher from the reporter dye, restoring fluorescence (indicated by *asterisk* and *bright green color* of reporter dye)

pared to the ABI system. At this time, there is less clinical experience for clinical infectious diseases testing for the remainder of instruments shown in Table 3.

Two of the real-time systems manufactured by Applied Biosystems, the GeneAmp 5700 and PRISM 7700 as well as and the LightCycler system (Roche Molecular Biochemicals) have been coupled with automated nucleic acid extraction instruments: ABI PRISM 6700 (Applied Biosystems) and MagNA Pure LC (Roche Molecular Biochemicals), respectively. As a result, the entire process of nucleic acid extraction, amplification, and detection is automated.

Advantages of Real-Time PCR

Compared to conventional PCR-based formats, real-time PCR platforms offer several important advantages (Table 4):

1. As mentioned previously, real-time PCR-based formats are performed using closed systems; therefore, compared to conventional PCR test methods, the potential for amplicon contamination is low.

Table 4. Advantages of real-time versus conventional PCR detection methods

	Real-time PCR	Conventional PCR
Sample preparation (lysis of organism, isolation, and purification of nucleic acids)	Specific instruments available for integration with ABI Gene-Amp 5700 and ABI Prism 7700 (ABI Prism 6700) and LightCycler (MagNA Pure LC)	Specific instrument available for integration with AMPLICOR-COBAS (AmpliPrep); other automated instruments for lysis/extraction of nucleic acid can be used for manual PCR platforms (see Table 8)
Amplification and detection	Integrated, simultaneous, closed system	Automated with COBAS, otherwise multi-step, open systems
Detection formats	Fluorophores	Gels, radioisotopes, digoxigenin, biotin, enzyme assays
Multiplex	Yes	Yes and no
TAT (analytic)	20 min to 2.0 h	4 h to 48 h
Amplicon contamination rate	Low	Moderate to high[a]
Assay capabilities	Same platform for qualitative, quantitative, and mutation assays	Different platforms for qualitative, quantitative, and mutation assays
Quantitation range	5–6 logs	2–3 logs

[a] Contamination potential is low if amplicons are sterilized as with AMPLICOR-COBAS system.

2. The time required to complete real-time assays, analytical turnaround time (TAT), is much shorter than that required for conventional PCR methods, including the automated COBAS system.
3. The assay capabilities are greater for the real-time instruments. Qualitative, quantitative, mutation and multiplex assays can be performed with the same instrument. In comparison, different platforms are frequently required to perform these different analyses using conventional techniques.
4. The quantitation range for real-time PCR methods is significantly greater (5–6 logs) than conventional PCR methods (2–3 logs).

Most of the remainder of the discussion will focus on the LightCycler real-time platform. Because the LightCycler was the first commercially available real-time PCR instrument which permitted real-time data analysis, a larger number of clinical studies have been reported for this instrument.

The LightCycler is much faster than conventional PCR detection methods (including the automated COBAS instrument) and the first available real-time instruments, the ABI GeneAmp 5700 and PRISM 7700. The entire amplification and detection process is automated and can be accomplished in as little as 30–40 min. Therefore same-day testing is easily accommodated. As mentioned above in "Real-time PCR Testing Formats Used in the Clinical Microbiology Laboratory", continuous or near continuous flow of samples can occur and the technologist is able to assess amplicon production at the end of each PCR cycle. With each run requiring less than an hour, new samples can be added in rapid succession.

Preanalytical processing of samples (lysis of organisms, nucleic acid extraction and purification) for the LightCycler can also be accomplished in a completely automated process using the MagNA Pure LC Instrument (Roche Molecular Biochemicals). The MagNA Pure LC instrument also loads the extracted nucleic acid into cuvettes and inserts these cuvettes into a rotor. Technologists remove the rotor and place it into a centrifuge and after centrifugation place the rotor into the LightCycler. These are the only manual steps in the process. In summary, labor requirements for nucleic acid extraction and analysis are substantially reduced when the MagNA Pure and LightCycler instruments are used together. In our experience, the combination of these two instruments represents the greatest advancement to date in real-time PCR technology.

Another feature of the LightCycler is that additional instruments can be added with expanding test repertoires. Finally, the size of LightCycler instruments (footprint) is relatively small. Footprints for the LightCycler are similar to conventional thermocyclers and therefore permit integration of this instrumentation into smaller laboratories.

Applications of Real-Time PCR for Diagnostic Testing in the Clinical Microbiology Laboratory

The versatility of the LightCycler for the diagnosis of infectious disease is demonstrated by the following broad applications. This platform can replace conventional PCR-based assays which have been used for identification or quantification of microorganisms that are nonculturable such as *Tropheryma whippeli*, the agent of Whipple's Disease, or grow very poorly, such as *Bordetella pertussis*. The LightCycler method can also replace conventional PCR-based assays that have been used to identify or quantify microorganisms that can be cultivated but require prolonged times for growth such as *Mycobacterium tuberculosis*.

For organisms that are easily cultured and grow relatively fast and for which conventional PCR has not been used for identification, the LightCycler real-time PCR may be particularly useful. For example, the current gold standard diagnostic approach for detecting Group B Streptococcus (*Streptococcus agalactiae*) in anal/vaginal swabs from pregnant women is culture. Colonization of the maternal genital tract with Group B Streptococcus is associated with risk of serious neonatal disease. Although these organisms grow relatively rapidly (1–2 days), a more rapid diagnosis is preferable so that antimicrobial prophylaxis can be provided as quickly as possible. Studies have demonstrated that the earlier chemoprophylaxis is initiated, the more likely it will be effective [1].

Rapid real-time PCR methods such as the LightCycler also have the potential to replace direct (antigen) immunoassays or direct microscopy. The enteric parasite *Giardia lamblia* can be detected by either direct immunoassay (microtiter enzyme immunoassay) or direct microscopy. Replacement of both of these assays by real-time PCR should significantly decrease labor requirements and improve analytical turnaround time.

The benefits of rapid turnaround time for antimicrobial susceptibility results and rapid detection of respiratory viruses have been demonstrated by Doern and colleagues [2] and Barenfanger and colleagues, respectively [3]. Both of these research groups have clearly shown that significant cost savings can be realized as the result of more rapid microbiology testing. As an example, control of vancomycin-resistant enterococci (VRE) in health care facilities has been demonstrated to be dependent on an active surveillance (culture-based) program [4]. Detection of VRE directly from stool specimens, as is possible using the LightCycler instrument, could significantly impact on the nosocomial spread of this pathogen. Current culture-based methods require at least 24 h to detect VRE. Preliminary studies conducted in our laboratory show that VRE can be detected directly from stool samples in less than 1 h by using a LightCycler assay.

Finally, the utility of the LightCycler method is further demonstrated from the biosafety perspective. By using this nonculture-based method, biohazard risks should be considerably lessened for the technologist. This attribute has particular relevance to organisms such as *M. tuberculosis* or organisms that might be used for bioterrorism attacks such as *Bacillus anthracis*.

Specific examples of microorganisms for which LightCycler assays may have particular utility for the reasons stated above are provided in Table 5. This is not meant to be a complete list but represents targeted assays which we have developed or plan to develop in the Mayo Clinic Microbiology Laboratory. Furthermore, our laboratory performs both common tests and esoteric tests; therefore, some tests listed may be more suitable for a reference laboratory.

When evaluating new testing platforms such as the LightCycler instrument with conventional technology for our laboratory, we compare specific parameters. These include analytical and clinical test performance as well as turnaround time (TAT) for results and cost.

Specific measures of analytical test performance include sensitivity and reproducibility. For this assessment, the ability of the LightCycler instrument to detect known quantities of target nucleic acid is evaluated. For determining clinical sensitivity and specificity, the results for the LightCycler assay are compared to the results for a conventional method, usually culture-based. Discordant results are reconciled by a review of the patient's medical history. One other assessment of clinical specimens is the determination of the frequency of inhibition of PCR by unknown substances in the sample. To screen for inhibition, a positive internal control is included in the PCR reaction mix. The control is a plasmid construct into which is inserted the target sequence to which the assay is directed.

Turnaround times are assessed as the total amount of time from the beginning of the analysis to the issuance of the final report. Costs are assessed for reagents and instruments, laboratory personnel, and laboratory space. Bedside cost savings are not always easily calculated but ideally the assessment of total savings for real-time PCR platforms should include this financial data. We have observed some turnaround times for real-time PCR results that are substantially lower than results for conventional testing methods and cost savings are therefore obvious. For example, the detection of *M. tuberculosis* and determination of isoniazid resistance in a sputum specimen in approximately 40 min compared to a minimum of 2 weeks for the same analysis using culture-based methods will permit appropriate directed therapy much sooner.

Table 5. Potential LightCycler assays for infectious agents

Organism	Nonculturable or poorly culturable	Culturable, but slow-growing	Culturable, relatively fast growth, but more rapid diagnosis preferred	Replacement for direct immunoassay or direct microscopy	Susceptibility loci (genes or mutations)	Biohazard
Bacteria						
Group A Streptococcus			√	√		
Group B Streptococcus			√	√		
Bordetella pertussis	√			√		
Bordetella parapertussis	√			√		
Bartonella henselae	√					
Bartonella quintana	√					
Ehrlichia chaffeensis	√			√		
Human granulocytic agent (HgE)	√			√		
Ehrlichia ewengii	√			√		
Legionella genus assay (pulmonary secretions)	√	√		√		
Legionella pneumophila species-specific assay (pulmonary secretions)		√		√		
Bacillus anthracis			√			√
Legionella genus (urine)				√		
Legionella pneumophila (urine)				√		
Screen for pathogens in positive blood cultures: *Staphylococcus aureus, Escherichia coli, Staphylococcus* species, coagulase-negative, *Enterococcus* species, *Candida albicans, Pseudomonas aeruginosa, Klebsiella pneumoniae, Streptococcus* species, viridans group, *Enterobacter cloacae, Streptococcus pneumoniae*			√			
Screen for bacterial pathogens in CSF: *S. pneumoniae, Neisseria meningitidis, Haemophilus influenzae*, Group B Streptococcus			√	√		
Screen for bacterial pathogens in CSF: *Listeria monocytogenes*			√			
Screen of platelet concentrates for bacterial contamination (16 S rDNA)			√			
E. coli O157 and other Shiga-toxin-producing *E. coli* (stool)			√	√		
E. coli O157 (urine)				√		
Salmonella spp.			√	√		
Shigella spp.			√	√		
Campylobacter jejuni			√	√		
Yersinia enterocolitica			√			
Clostridium difficile toxin A				√		

Table 5. *Continued.*

Organism	Noncul-turable or poorly culturable	Cultur-able, but slow-growing	Culturable, relatively fast growth, but more rapid diagnosis preferred	Replacement for direct immuno-assay or direct microscopy	Suscepti-bility loci (genes or mutations)	Bio-haz-ard
Clostridium difficile toxin B				√		
H. pylori (including *cag* gene), stool		√		√		
Borrelia burgdorferi (Lyme)	√					
Mycoplasma genus assay			√			
Mycoplasma pneumoniae species-specific assay			√			
Chlamydia pneumoniae	√					
M. tuberculosis – identification		√		√		√
Screen for *M. tuberculosis* in tissue		√		√		√
Mycobacterium genus screen (sputum, tissue)		√		√		√
Rapid-growing Mycobacterium screen		√				
Treponema pallidum (syphilis)	√			√		
Tropheryma whippelii (agent of Whipple's disease)	√			√		
Methicillin (oxacillin)-resistant (*mecA* gene) staphylococci					√	
Vancomycin-resistant (*vanA, vanB* genes) enterococci (stool)					√	
Extended-spectrum β-lactamase (ESBL) genes, gram-negative bacteria (genes which are derivatives of TEM, SHV, and OXA-type β-lactamases)					√	
M. tuberculosis – *katG* S315 T (isoniazid resistance)					√	√
M. tuberculosis – *rpoβ* (rifampin resistance)					√	√
speA gene for Group A Streptococcus (encodes pyrogenic exotoxin A)				√		
speC gene for Group A Streptococcus (encodes pyrogenic exotoxin C)				√		
Staphylococcal toxic shock syndrome toxin 1 (TSST-1) gene				√		
Viruses						
Herpes simplex virus 1 CSF	√					
Genital, dermal			√	√		
Herpes simplex virus 2 CSF	√					
Genital, dermal			√	√		
Varicella-zoster virus,	√					
CSF Dermal			√	√		
Cytomegalovirus (quantitative assay blood)		√		√		

Table 5. *Continued.*

Organism	Nonculturable or poorly culturable	Culturable, but slow-growing	Culturable, relatively fast growth, but more rapid diagnosis preferred	Replacement for direct immunoassay or direct microscopy	Susceptibility loci (genes or mutations)	Biohazard
CMV qualitative,						
urine and samples other than blood		√				
CSF	√					
EBV (Epstein-Barr virus) (quantitative assay), blood	√					
EBV (qualitative assay), CSF	√					
Human herpesvirus-6 (quantitative assay)	√					
BK virus (quantitative)	√					
JC virus – CSF	√					
RSV (respiratory syncytial virus)		√		√		
Arbovirus	√					
Flavivirus	√					
Rabies virus	√			√		
Influenza virus A		√		√		
Influenza virus B		√		√		
Enterovirus (generic)			√			
Adenovirus (other than serotypes 40, 41)		√				
Adenovirus 40, 41 (stool pathogens)	√					
Parvovirus	√					
Rotavirus (stool)	√			√		
Fungi						
Histoplasma capsulatum						
Specimens other than urine		√		√		
Urine		√		√		
Pneumocystis carinii	√			√		
Blastomyces dermatitidis		√		√		
Coccidioides immitis		√		√		√
Penicillium marnefei		√				
Fungemia screen: *H. capsulatum*		√				
Fungemia screen: *Coccidioides immitis*		√		√		√
Fungemia screen: *Cryptococcus neoformans*		√		√		
Screen for fungal pathogens in CSF: *Cryptococcus neoformans, Coccidioides immitis*		√		√		
Screen for fungi in tissue		√		√		√

Table 5. *Continued.*

Organism	Noncul-turable or poorly culturable	Cultur-able, but slow-growing	Culturable, relative-ly fast growth, but more rapid diagno-sis preferred	Replacement for direct immuno-assay or direct microscopy	Suscepti-bility loci (genes or mutations)	Bio-haz-ard
Parasites						
Giardia lamblia				√		
Entamoeba histolytica				√		
Cryptosporidium sp.				√		
Babesia microti (blood)				√		
Microsporidium sp.				√		
Plasmodium falciparum				√		
Plasmodium vivax				√		
Plasmodium ovale				√		
Plasmodium malariae				√		
Toxoplasma gondii				√		

Table 6. Comparison of LightCycler assays to gold standard (culture-based) assays for selected pathogens – Mayo Clinic experience

Organism	Publication reference number	Increase in sensitivity for LightCycler compared to culture	Analytical TAT[a]	
			Light-Cycler	Culture-based gold standard
Group A Streptococcus	5	7%	30 min	2 days
Legionella sp. (BAL)[b]	6	Equal[c]	45 min	14 days
Bordetella pertussis	7	219%	50 min	7 days
Varicella-zoster (skin)	8	91%	45 min	5 days
Herpes simplex (skin, genital)	9, 10	23%	45 min	5 days
Cytomegalovirus (urine)	11	88%	45 min	24 h (shell vial only)

[a] TAT, turnaround time.
[b] BAL, bronchoalveolar lavage.
[c] Only archived culture-positive samples available for this comparative study.

Table 6 shows sensitivity and analytical TAT data for clinical trials conducted at the Mayo Clinical Microbiology Laboratory, which compared LightCycler assays to standard culture-based methods for detecting pathogens in human specimens [5–11]. As can be seen, except for *Legionella* spp., the LightCycler assays had remarkably higher clinical sensitivities than standard culture methods. The LightCycler assay was equivalent in sensitivity to culture for detection of *Legionella* spp. in bronchoalveolar lavage (BAL) specimens, but only archived specimens were available for study. As can also be observed in the table, a dramatic decrease in analytical time was noted for the LightCycler assays compared to the culture-based assays.

We have also recently applied LightCycler for quantifying Epstein-Barr virus in human peripheral blood leukocytes [12] and for detecting mutations associated with antimicrobial drug resistance. Figure 4 shows one example of this latter application. We and others have demonstrated that a mutation, *S315T* in the catalase (*katG*) gene of *Mycobacterium tuberculosis* is associated with high-level isoniazid resistance [13]. Isoniazid is the most frequently prescribed drug for treatment of and prophylaxis against infection with *M. tuberculosis*. As shown in Fig. 4, by using a melting curve analysis of FRET probes, one can easily distinguish the wild-type (WT) *katG* from the mutated (*S315T*) *katG*.

Table 7 provides estimates for reagent and/or instrument costs and personnel time required for detection of Group A Streptococcus (Streptococci parenthesis)

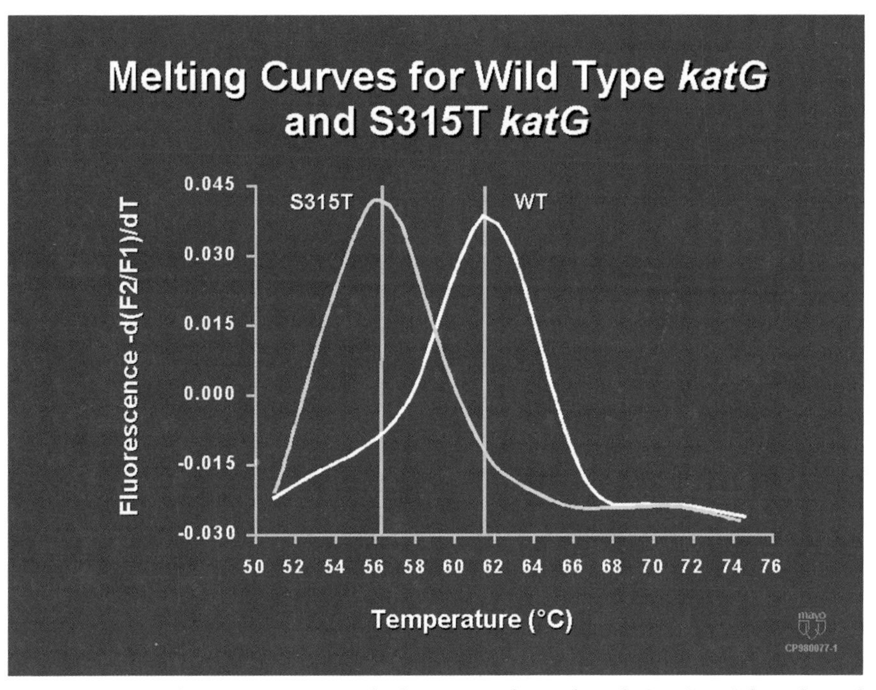

Fig. 4. Detection of *S315T* mutation in the *katG* gene of *M. tuberculosis* using LightCycler and melting curve analysis. A FRET probe will "melt" (separate) from its complementary target DNA strand at a lower temperature if there is a mutation in the target DNA strand. This mutation results in a nucleotide "mismatch" of the probe and target DNA at which point annealing of probe and target does not occur. As a result, as the temperature is increased, the probe and mutated target DNA will denature (melt) sooner than wild-type target DNA, where no mismatch occurs. In this example, the mutated target *katG* (*S315T*) DNA and its complementary FRET probe melt at a lower temperature than the wild-type (WT) target DNA. The *S315T* mutation is associated with high-level isoniazid resistance

Table 7. Comparison of LightCycler assay to gold standard assay for Group A Streptococcus

	Personnel time/test[a]	Cost/test
LightCycler	~3 min	\$1-\$2[b]
Rapid antigen/culture	~7 min	\$2-\$3[c]

[a] Based on workload recording standards as provided by the National Committee for Clinical Laboratory Standards (NCCLS) (15).
[b] Cost for reagents and depreciation of LightCycler instrument over 3 years; royalty for PCR not included.
[c] Cost for reagents.

from throat swabs for LightCycler compared to rapid antigen detection and culture. In our practice, in accordance with practice guidelines provided by the Infectious Diseases Society of America (IDSA) [14], all rapid antigen tests that are negative are backed-up by culture, i.e., if a rapid antigen test is negative, a confirmatory culture is performed. This protocol is followed as rapid antigen tests, which, although highly specific, lack sensitivity. As shown in Table 7, the cost per test for the LightCycler Group A Streptococcus assay was about half of that for the combined rapid antigen and culture analysis. This cost analysis does not include PCR royalty costs which are assessed on total test charges to the patient (usually ~15%). Even if this additional cost is considered, the total cost for performing the LightCycler test is considerably less than the cost for the rapid antigen test and culture.

As also indicated in Table 7, the LightCycler Group A Streptococcus assay required less than half the time of the rapid antigen/culture assay to perform. The amount of time required for each of these assays was based on workload recording standards as provided by the National Committee for Clinical Laboratory Standards (NCCLS) [15]. With the rapidity with which LightCycler assays can be performed, results can be relayed to healthcare providers and patients the same day the test is ordered. At Mayo Clinic, ambulatory patients with pharyngitis can have throat swabs performed by a nurse practitioner without a physician visit. The results from analyses are entered electronically into a computerized telephone call-back system. To access their results, patients call an 800 number and listen to a recorded message. The results of the test are provided and the patient is notified that a prescription can be picked up at a pharmacy preselected by the patient. Such rapid turnaround and convenient reporting of results is welcomed by our patients and should impact significantly on the appropriate and expedient delivery of antimicrobics for streptococcal pharyngitis.

Challenges of the LightCycler in the Clinical Microbiology Laboratory

Figure 1 shows the steps involved in testing human specimens from specimen collection and transport to the final reporting of results. The development of real-time PCR assays has effectively combined the third and fourth steps, PCR and detection, into a single highly efficient step. Indeed, in our experience, the time required for the actual analysis of a specimen by the LightCycler instrument in many cases is considerably less than the time required to collect and transport the specimen to the laboratory and in some cases is even less than the time required to report the results.

Collection of a suitable specimen, the first step shown in Fig. 5, is the most important step in the testing process. If the quality of the specimen is poor, the chances of detecting, or quantifying pathogens, identifying virulence genes (e.g., toxin genes), or detecting genes or mutations in pathogens associated with antimicrobial resistance, may be severely limited. The proverbial expression "garbage in, garbage out" applies not only to traditional culture-based assays but as well to PCR-based assays. Over the years, standards have been developed to assess specimen quality for some conventional culture-based testing methods. One example is the microscopic screening of Gram stains of sputum specimens to determine the suitability of specimens for bacterial culture when lower respiratory tract infection is suspected. Murray and Washington [16] have demonstrated that the most suitable sputum specimens for bacterial culture should have fewer than ten squamous epithelial cells observed per low-power microscopic field, indicating that the specimen is not excessively contaminated with oropharyngeal secretions. These investigators also noted that if white cells are present with less than ten squamous epithelial cells per low-power field, it is even more convincing that an adequate specimen of deep pulmonary secretions has been obtained and that a lower respiratory bacterial infection is present. Similar standards, for exam-

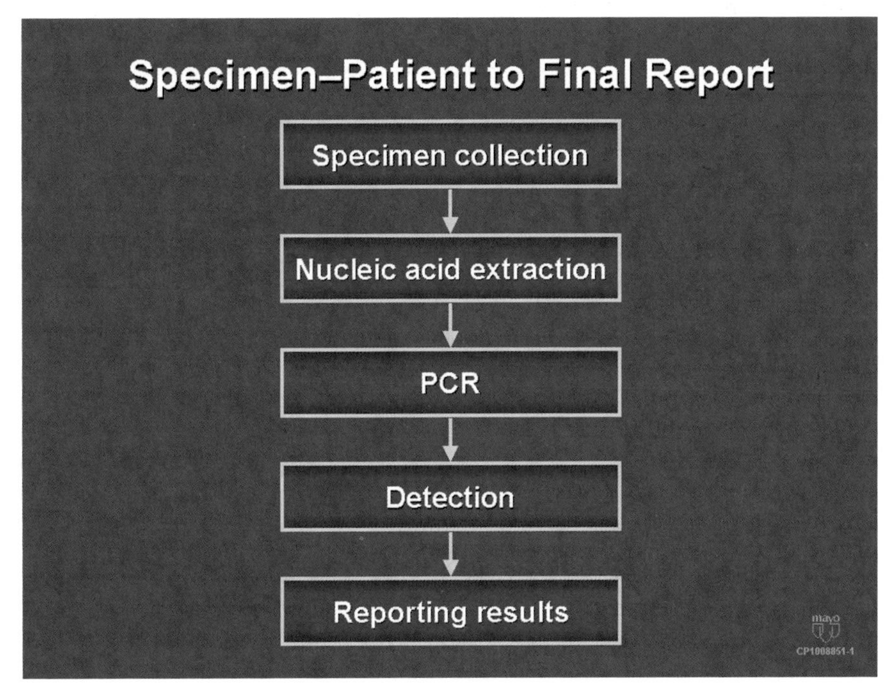

Fig. 5. Specimen–patient to final report

ple, screening for DNA from human cells, may be useful for real-time PCR testing platforms to assure that high quality specimens are evaluated.

Calliendo and colleagues (personal communication) have made some attempt to define the quality of throat swabs for *Mycoplasma pneumoniae* detection by conventional PCR analysis. They recommend screening throat swab specimens for the presence of human DNA. If human DNA is not amplified, the implication is that the throat epithelial cells were not adequately swabbed (sampled) and therefore the specimen is not suitable for detection of *M. pneumoniae* by PCR.

Of additional interest, the type of specimen requested to diagnose a specific infection may be different for real-time PCR testing methods vs culture-based methods. As an example, for many years a nasopharyngeal aspirate has been recommended over a throat swab as the optimal specimen for recovering *Bordetella pertussis* by culture. As indicated in Table 6, we have noted for samples collected from the nasopharynx that the sensitivity of a real-time PCR assay for *B. pertussis* detection is 219% greater than culture. This finding is similar to that reported by Loeffelholz and colleagues [17] who noted a roughly 250% increase in sensitivity for conventional PCR vs culture for detecting *B. pertussis*. It is possible that because real-time PCR is so much more sensitive for detecting this organism as compared to culture, that pharyngeal swabs may be suitable specimens for evaluation by real-time PCR. If so, patients would experience considerably less discomfort as the collection of a nasopharyngeal aspirate requires insertion of a swab or catheter through the anterior nares to the posterior nasopharynx. However, before throat swabs can be recommended for real-time PCR, prospective studies must be undertaken to compare the sensitivity of detection of *B. pertussis* from nasopharyngeal specimens compared with throat specimens for this method. Another issue is what type of swab should be used for PCR. Certain swab fibers or medium used to transport swabs may be inhibitory to PCR. Studies evaluating the inhibitory effects of various commercial swabs and transport media on PCR must also be conducted.

The transport of the specimens to the laboratory may present challenges, especially if RNA is the target nucleic acid for the real-time PCR assay. RNA viruses are labile to RNase enzymes which may arise from human cells in specimens. These specimens should be transported on ice to inhibit RNase activity. In contrast, DNA-containing organisms may be shipped at room temperature.

In summary, for real-time PCR testing platforms, studies are required to determine the optimal specimen type and specimen collection (transport) devices for identification of a variety of human pathogens. Specimen type and collection guidelines for culture-based, antigen-based, or serological-based detection methods may not apply to real-time PCR detection methods.

The second major step in the testing process outlined in Fig. 1 involves preanalytical processing. Our research group is now focusing on this very important step. The efficient extraction and concentration of target nucleic acid in samples may be of particular importance, especially in conditions where very low quantities of a pathogen are present in a specimen (e.g., central nervous system infections). Additionally, monitoring for, as well as removal of substances that may inhibit the PCR reaction, is an important quality measure. If significant amounts of PCR inhibitors are present, even relatively high quantities of nucleic acid target may not be detected. Therefore, the incorporation of an internal or inhibition control as discussed previously (usually a plasmid in which the target nucleic acid sequence is spliced) in the PCR mix is essential to assure that the PCR is working.

Recently, semiautomated and automated platforms have been developed for extraction of nucleic acid from human samples. As previously indicated, the MagNA Pure LC instrument has been developed for use specifically with the LightCycler instrument and both of these instruments are completely automated.

The principles of nucleic acid extraction for commonly used conventional manual and automated and semiautomated methods are shown in Table 8. Of note, many of these platforms have been evaluated and adapted for nucleic acid extraction from blood, plasma, or serum. The utility of these methods for extracting nucleic acids from swab specimens, pulmonary secretions, synovial or spinal fluid, urine, feces and tissue has not been adequately studied. Furthermore, modifications in these procedures may be required to optimize nucleic acid extraction from these sources. A recent study at our institution has demonstrated no statistically significant difference in the ability to extract herpes simplex DNA from the same specimen for either the IsoQuick manual method or the semiautomated Qiagen method or automated MagNA Pure method [18]. For this study, herpes simplex DNA was detected using the LightCycler instrument.

To assure the universal adaptability, ease of performance and quality of real-time PCR assays, standard protocols should be developed for their use. In this regard, development of analyte specific reagents (ASRs) or FDA-approved kits by vendors of real-time PCR is encouraged. Proficiency testing of unknowns

Table 8. Examples of commonly used manual and automated nucleic acid extraction platforms

Name of method/ instrument	Company	Chemistry	Comments
IsoQuick	Orca Research, Inc. (Bothell, WA)	Proprietary	Manual
Biorobot 9604	Qiagen, Inc. (Chatsworth, CA)	Lysis in guanidinium buffer, nucleic acid capture to silica-gel membrane	Robotic workstation with manual centrifugation steps
ABI PRISM 6700	Applied Biosystems (Foster City, CA)	Proprietary	Designed for integration with ABI Gene-Amp 5700 and ABI PRISM 7700 instruments
NucliSense Extractor	Organon Teknika Corporation (Durham, NC)	Lysis in guanidinium buffers; nucleic acid bound to silicon dioxide particles	Designed for integration with nucleic acid sequence-based analysis (NASBA)
MagNA Pure LC	Roche Molecular Biochemicals	Lysis in guanidinium buffer; nucleic acid capture with magneticsilica particle capture	Designed for integration with LightCycler instrument
COBAS AmpliPrep	Roche Diagnostics Corporation	Lysis in guanidinium buffer; nucleic acid capture with strepta-vidin-coated magnetic beads and biotinylated probes	Designed for integration with the COBAS instrument

provided by manufacturers of ASRS/kits and objective third-party agencies such as the College of American Pathologists (CAP), Northfield, IL, will contribute to quality assurance. Finally, to avoid any risk of amplicon contamination, we recommend that all real-time PCR assays incorporate the amplicon sterilization step. In our laboratory, all LightCycler assays include the incorporation of deoxyuridine (dUTP) instead of deoxythymidine (dTTP) in amplicons; any amplicons that may contaminate subsequent assays are cleaved by uracil-*N*-glycosylase (UNG).

Conclusions

The extensive applications of LightCycler real-time PCR assays for the Clinical Microbiology Laboratory as evidenced by the preceding discussion make it clear that real-time PCR formats such as LightCycler will significantly change the way we identify or quantify pathogens and determine their virulence factors or antimicrobial resistance.

Real-time PCR has immediate and important implications for diagnostic testing in the clinical microbiology laboratory. In our hands, real-time PCR, using the LightCycler instrument has outstanding performance characteristics; the sensitivities for detection of microorganisms frequently exceeds standard culture-based assays. The LightCycler instrument permits a much shorter turnaround time for results, is easy to perform, and relatively inexpensive compared with culture-based assays. In many cases, assays performed using LightCycler will replace culture as the gold standard for identification of microorganisms. Development of ASR kits by vendors of real-time PCR technology and standardization of testing formats are necessary to assure quality of testing.

References

1. Edwards MS, Baker CJ (2000) *Streptococcus agalactiae* (Group B Streptococcus). In: Mandell GL, Bennett JE, Dolin R (eds) Principles and practice of infectious diseases, 2nd edn. Churchill Livingstone, Philadelphia, pp 2156–2167
2. Doern GV, Vautour R, Gaudet M (1994) Clinical impact of rapid in vitro susceptibility testing and bacterial identification. J Clin Microbiol 32:1757–1762
3. Barenfanger J, Drake C, Leon N, Mueller T, Trout (2000) Clinical and financial benefits of rapid detection of respiratory viruses. J Clin Microbiol 38:2824–2828
4. Ostrowsky BE, Trick WE, Sohn AH, Quirk SB, Holt S, Carson LA, Hill BC, Arduino MJ, Kuehnert MJ, Jarvis WR (2001) Control of vancomycin-resistant enterococcus in health care facilities in a region. N Eng J Med 344:1427–1434
5. Uhl JR, Vetter EA, Cockerill FR III (2001) Detection of Group A Streptococcus from throat swabs using rapid homogeneous PCR (abstract). Abstracts of the 101st General Meeting of the American Society for Microbiology. No. C-21. Orlando, Florida, May 20–24, 2001
6. Hayden RT, Uhl JR, Qian X, Hopkins MK, Aubry MC, Limper AH, Lloyd RV, Cockerill FR (2001) Direct Detection of *Legionella* species from bronchoalveolar lavage and open lung biopsy specimens: a comparison of LightCycler PCR, in situ hybridization, direct fluorescence antigen detection and culture. J Clin Microbiol 39:2618–2626

7. Sloan LM, Hopkins MK, Mitchell PS, Vetter EA, Rosenblatt JE, Cockerill FT, Patel R (2001) A Multiplex LightCycler polymerase chain reaction assay for the detection and differentiation of *Bordetella pertussis* and *Bordetella parapertussis* in nasopharyngeal specimens. J Clin Microbiol
8. Espy MJ, Teo R, Ross K, Svien KA, Wold AD, Uhl JR, Smith TF (2000) Diagnosis of varicella-zoster virus infections in the clinical laboratory by LightCycler PCR. J Clin Microbiol 38:3187–3189
9. Espy MJ, Uhl JR, Mitchell PS, Thorvilson JN, Svien KA, Wold AD, Smith TF (2000) Diagnosis of herpes simplex virus infections in the clinical laboratory by LightCycler PCR. J Clin Microbiol 38:795–799
10. Espy MJ, Ross TK, Teo R, Svien KA, Wold AD, Uhl JR, Smith TF (2000) Evaluation of LightCycler PCR for implementation of laboratory diagnosis of herpes simplex virus infections. J Clin Microbiol 38:3116–3118
11. Espy MJ, Smith TF (2000) Detection of cytomegalovirus (CMV) in clinical specimens by LightCycler PCR (abstract). Abstracts of the 100th General Meeting of the American Society for Microbiology, No. C-62. Los Angeles, CA, May 21–25, 2000
12. Espy MJ, Patel R, Paya CV, Smith TF (2001) Quantification of Epstein-Barr virus (EBV) viral load in transplant patients by LightCycler PCR (abstract). Abstracts of the 101st General Meeting of the American Society for Microbiology, No. C-148. Orlando, Florida, May 20–24, 2001
13. Uhl JR, Sandhu GS, Kline BC, Cockerill FR III (1996) PCR-RFLP detection of point mutations in the catalase-peroxidase gene (*katG*) of *Mycobacterium tuberculosis* associated with isoniazid resistance. In: Persing DH (ed) PCR Protocols for emerging infectious diseases. American Society for Microbiology, Washington, DC
14. Bisno AL, Gerber MA, Gwaltney JM, Kaplan EL, Schwartz RH (1997) Diagnosis and management of group A streptococcal pharyngitis: a practice guideline. Clin Infect Dis 25:574–583
15. Basic cost accounting for clinical services; approved guidelines. November 1998. National Committee for Clinical Laboratory Standards, GP11-A, Vol. 18, Wayne, PA
16. Murray PR, Washington JA II (1975) Microscopic and bacteriologic analysis of expectorated sputum. Mayo Clin Proc 50:339–344
17. Loeffelholz MJ, Thompson CJ, Long KS, Gilchrist MJR (1999) Comparison of PCR, culture, and direct fluorescent-antibody testing for detection of *Bordetella pertussis*. J Clin Microbiol 37:2872–2876
18. Espy MJ, Rys PN, Wold AD, Uhl JR, Sloan LM, Jenkins GD, Ilstrup DM, Cockerill FR III, Patel R, Rosenblatt JE, Smith TF (2001) Detection of herpes simplex virus DNA in genital and dermal specimens by LightCycler PCR after extraction by IsoQuick, MagNA Pure, and BioRobot 9604. J Clin Microbiol 39:2233–2236

Bacteriology I

Rapid Detection and Simultaneous Differentiation of *Bordetella pertussis* and *Bordetella parapertussis* in Clinical Specimens by LightCycler PCR 31
Udo Reischl, Katrin Kösters, Birgit Leppmeier, Hans-Jörg Linde, Norbert Lehn

Rapid Detection and Simultaneous Differentiation of *Legionella* spp. and *L. pneumophila* in Potable Water Samples and Respiratory Specimens by LightCycler PCR 45
Nele Wellinghausen, Olfert Landt, Udo Reischl

Homogeneous Assays for the Rapid PCR detection of *Burkholderia pseudomallei* 16s rRNA gene on a Real-Time Fluorometric Capillary Thermocycler 59
Eric P-H Yap, Soon-Meng Ang, Shirley G-K Seah, Seng-Meng Phang

Rapid Detection of Toxigenic *Corynebacterium diphtheriae* by LightCycler PCR 71
Udo Reischl, Elena Samoilovich, Valentina Kolodkina, Norbert Lehn, Hans-Jörg Linde

Rapid Detection of Resistance Associated Mutations in *Mycobacterium tuberculosis* by LightCycler PCR 83
Maria J. Torres, Antonio Criado, Jose C. Palomares, Javier Aznar

Duplex LightCycler PCR Assay for the Rapid Detection of Methicillin-Resistant *Staphylococcus aureus* and Simultaneous Species Confirmation 93
Udo Reischl, Hans-Jörg Linde, Birgit Leppmeier, Norbert Lehn

Rapid Detection of Group B Streptococci Using the LightCycler Instrument 107
Danbing Ke, Christian Ménard, François J. Picard, Michel G. Bergeron

Rapid Detection and Quantification of *Chlamydia trachomatis* in Clinical Specimens by LightCycler PCR . . . 115
Heidi Wood, Udo Reischl, Rosanna W. Peeling

Quantification of *Toxoplasma gondii* in Amniotic Fluid by Rapid Cycle Real-Time PCR . . . 133
François Delhommeau, François Forestier

Genospecies-Specific Melting Temperature of the *recA* PCR Product for the Detection of *Borrelia burgdorferi* sensu lato and Differentiation of *Borrelia garinii* from *Borrelia afzelii* and *Borrelia burgdorferi* sensu stricto . . . 139
Johanna Mäkinen, Qiushui He, Matti K. Viljanen

Rapid and Specific Detection of *Coxiella burnetii* by LightCycler PCR . . . 149
Markus Stemmler, Hermann Meyer

Molecular Identification of *Campylobacter coli* and *Campylobacter jejuni* Using Automated DNA Extraction and Real-Time PCR with Species-Specific TaqMan Probes . . . 155
Lynn A. Cooper, Luke T. Daum, Robert Roudabush, Kenton L. Lohman

Rapid and Specific Detection of *Plesiomonas shigelloides* Directly from Stool by LightCycler PCR . . . 161
Jin Phang Loh, Eric Peng Huat Yap

Quantification of Thermostable Direct Hemolysin-Producing *Vibrio parahaemolyticus* from Foods Assumed to Cause Food Poisoning Using the LightCycler Instrument . . . 171
Yoshito Iwade, Akinori Yamauchi, Akira Sugiyama, Osamu Nakayama, Norichika H. Kumazawa

Rapid Detection and Simultaneous Differentiation of *Bordetella pertussis* and *Bordetella parapertussis* in Clinical Specimens by LightCycler PCR

Udo Reischl*, Katrin Kösters, Birgit Leppmeier,
Hans-Jörg Linde, and Norbert Lehn

Introduction

The bacterium *Bordetella pertussis* is the causative agent of whooping cough, which is an infectious disease occurring worldwide with a high incidence among young, unvaccinated infants. *B. pertussis* is transmitted by respiratory droplets and causes disease only in humans. Related species, which may also cause pertussis syndrome, are *B. parapertussis*, *B. holmesii*, and *B. bronchiseptica* (primarily infecting animals and less pathogenic in humans). The illness caused by *B. parapertussis* is usually milder than that caused by *B. pertussis*, although severe cases have been reported [1] and simultaneous infections with *B. pertussis* and *B. parapertussis* have also been observed in clinical practice. Therefore, a rapid and reliable diagnostic method which detects and differentiates between *B. pertussis* and *B. parapertussis* organisms is highly desirable [2, 3]. The striking and unique presentation of classical pertussis does not usually present a clinical diagnostic dilemma. Atypical pertussis, however, which may occur with mild or absent symptoms in adults or previously vaccinated children, offers a greater diagnostic challenge to the clinician. It has been shown that atypical illness in adults is common, endemic, and usually unrecognized [4]. The epidemiological implications of unrecognized pertussis are that exposure of unimmunized infants to individuals with pertussis places them at high risk and that pertussis remains endemic in society. The recent increase of pertussis in the United States [5] and France [6] and the belief that other agents such as adenovirus, parainfluenza viruses, cytomegalovirus, *Mycoplasma pneumoniae*, or *Chlamydia pneumoniae* cause pertussis syndrome [7] underscore the need for rapid and accurate diagnostic methods to guide therapeutic and preventive interventions.

Although culture of *Bordetella* spp. is considered to be the gold standard for diagnosis of pertussis due to its high specificity, it is maximally sensitive only in the initial phases of disease and colonies have to be verified by agglutination and biochemical tests. Moreover, successful culture requires special media, incubation periods up to 7 days, and is highly dependent upon specimen collection and lab-

* Udo Reischl (✉) (e-mail: Udo.Reischl@klinik.uni-regensburg.de)
Institute of Medical Microbiology and Hygiene, University of Regensburg,
Franz-Josef-Strauß-Allee 11, 93053 Regensburg, Germany

oratory techniques. Diagnostic sensitivities below 60% are observed when nasopharyngeal secretions are obtained outside the early catarrhal stage of the illness from older or vaccinated persons, from persons treated with certain antibiotics, or in the case of prolonged sample transport.

Serology is considered highly sensitive and can provide rapid diagnosis for a patient with classical pertussis, but there are two major drawbacks: it cannot be used in the acute phase of the disease, and it can be difficult to differentiate between vaccine effects and a pertussis infection. Moreover, serological testing is considered less reliable in infants and, due to its poor specificity, it should not be used to detect atypical disease. Thus there is a need for more rapid diagnostic methods with high degrees of specificity and sensitivity, preferably for use early in infection.

Similar to other fastidious organisms which are significant by their presence even in the asymptomatic individual, nucleic acid amplification techniques, such as PCR, are the method of choice for direct detection of *B. pertussis* and *B. parapertussis* in clinical specimens [3, 8–13]. Potential *B. pertussis*-specific target regions include insertion sequences, repeat elements, the pertussis toxin (PT) promoter region, the adenylate cyclase gene, and the porin gene.

Here we describe the development of a sensitive and specific hybridization probe-based LightCycler assay and its validation for the detection of *B. pertussis* and *B. parapertussis* organisms in clinical specimens. *B. pertussis* repetitive insertion sequence *IS481* and *B. parapertussis* repetitive insertion sequence *IS1001* were selected as targets for the duplex PCR. Testing a collection of 208 respiratory specimens, cultured strains of *B. pertussis*, *B. parapertussis*, *B. holmesii*, *B. trematum*, and *B. bronchiseptica*, as well as 80 isolates of Gram-negative and Gram-positive bacterial species other than *Bordetella*, a good correlation of PCR results was obtained compared to clinical findings and, in part, to the results of conventional microbiological testing. However, a remarkable cross-reaction of *B. holmesii* was observed with the *IS481*-specific assay (see "Comments").

Materials

Equipment

LightCycler instrument (Roche Diagnostics, Mannheim, Germany)
LightCycler software vers. 3.3
Oligo Primer analysis software (Molecular Biology Insights, Inc., Cascade, CO, USA)

Reagents

Amplification Primers (Metabion, Munich, Germany)
Hybridization Probes (TIB MOLBIOL, Berlin, Germany)

The following reagents were purchased from Roche Diagnostics:
High-Pure PCR Template Purification Kit
LightCycler Fast Start DNA Master Hybridization Probes
LightCycler Control Kit DNA

Procedure

Preparation of Template DNA

Bacterial DNA was prepared, using the High Pure PCR Template Preparation Kit according to the manufacturer's instructions. Starting with nasopharyngeal swabs, throat swabs, or perinasal swabs, about half of the swab tip was cut off, suspended in 200 µl of tissue lysis buffer and 40 µl of proteinase K solution (20 mg/ml) and incubated at 55°C for at least 30 min. Because calcium alginate swabs are associated with severe inhibition of the subsequent PCR process, standard swabs without special additives and, whenever possible, without dipping into transport medium, should be requested. Starting with nasopharyngeal aspirates or bronchoalveolar lavages, aliquots with a volume of approximately 100 µl were suspended in 200 µl of tissue lysis buffer and 40 µl of proteinase K solution (20 mg/ml) and incubated at 55°C for at least 30 min. After complete disintegration of tissue pieces, which can be examined visually, 200 µl of binding buffer were added and a further incubation was performed at 70°C for 10 min. Subsequently, 100 µl of isopropanol was added and the mixture was applied to the High Pure spin column.

Starting with cultured bacteria, single colonies were suspended in 200 µl of PBS buffer, 15 µl of a lysozyme solution (10 mg/ml in Tris-HCl, pH 8.0) was added and incubated at 37°C for 10 min. After adding 200 µl of binding buffer and 40 µl of a proteinase K solution (20 mg/ml) with a further incubation at 70°C for 10 min, 100 µl of isopropanol was added and the mixture was applied to the High Pure spin column.

Following the centrifugation and wash steps, bacterial DNA was eluted with 200 µl of elution buffer and a 2-µl aliquot was directly transferred to PCR. The remainder was stored at –20°C for further experiments.

Primer and Hybridization Probe Design

A previously published primer pair [9] (see Table 1) was used for amplifying a segment within the *B. pertussis* repetitive insertion sequence *IS481*, which proved to be a very sensitive target. A second primer pair [3] (see Table 1) was selected to amplify a segment within the *B. parapertussis* repetitive insertion sequence *IS1001*. Based on the primer annealing sites, alignments were performed with all of the different *IS481* and *IS1001* sequence entries deposited in GeneBank and EMBL databases to recognize and consider single nucleotide ambiguities. The selection of candidate sequences for LightCycler hybridization probes was aided by the Oligo software in order to obtain comparable T_m, GC contents within the range of 35%–60%, and to avoid stable secondary structures, regions of significant self-complementarity, stretches of palindromic sequences, and close proximity between the primers and the probes. For the sequence-specific detection of *IS481* amplicons, a set of hybridization probes was selected (see Table 1). In contrast to a *B. pertussis*-specific LightCycler protocol, published in a previous volume of this book series [14], the hybridization probes are now positioned on the opposite strand of the target sequence (O. Landt, TibMolbiol, Berlin, Germany). The former probe contained six guanosine bases in line, which have now been converted into a stretch of cytosines. Oligonucleotides containing stretches of guanosine residues or very long stretches of purine residues are known to cause problems during the hybridization process, presumably due to their higher degree of hydrophobicity. Binding of such oligonucleotides is less specific, leading to broad melting curves

Table 1. Oligonucleotides

Bordetella pertussis Repetitive Insertion Sequence IS481 (GenBank Accession #M22031, X58488, and U07800)			
	Length	GC (%)	T_m (°C)
Primers			
GATTCAATAGGTTGTATGCATGGTT	25	36.0	61.1
TTCAGGCACACAAACTTGATGGGCG	24	52.0	67.6
Probes			
TCGCCAACCCCCCAGTTCACTCA-F	23	60.9	71.3
LCRed640-AGCCCGGCCGGATGAACACCC	21	71.4	72.1
Bordetella parapertussis Repetitive Insertion Sequence *IS1001* (GenBank Accession #X66858)			
Primers			
CACCGCCTACGAGTTGGAGAT	21	57.1	65.3
CCTCGACAATGCTGGTGTTCA	21	52.4	64.3
Probes			
GTTCTACCAAAGACCTGCCTGGGC-F	24	58.3	68.5
LCRed705-AGACAAGCCTGGAACCACTGGTAC	24	54.2	67.4

and lower fluorescent signals. This may also be due to, at least in part, their lower coupling efficiencies during the probe synthesis, leading to the presence of side products with single base deletions (O. Landt, personal communication). For the sequence-specific detection of *IS1001* amplicons, a set of hybridization probes was selected (see Table 1). Annealing sites of primers and hybridization probes within the corresponding GenBank sequences are depicted in Figs. 1 and 2.

Inhibition Control

The LightCycler Control Kit DNA was used for the detection of *Taq* DNA polymerase inhibitors, possibly present in DNA preparations from clinical samples. This kit provides a control DNA template, primers, and hybridization probes for amplifying and specific detection of the human β-globin gene. To identify even weak inhibition events, 300 pg of human genomic DNA was amplified in each case.

LightCycler PCR

The following master mix was used for amplification and hybridization probe-based detection of the *IS481*- and *IS1001*-specific amplicons:

	Volume [μl]	[Final]
LightCycler Fast Start DNA Master Hybridization Probes	2	1×
$MgCl_2$ stock solution (25 mM)	1.6	3 mM
Primers (10 μM each)	1+1+1+1	0.5 μM each
Hybridization probes (2 μM each)	2+2+2+2	0.2 μM each
H_2O (PCR grade)	2.4	
Total volume	18	

```
                         Primer >
PCR Prod  1        gattcaataggttgtatgcatggttcatccgaaccggatttgagaaactggaaatcgcca 60
M22031    12       ............................................................ 71
X58488    227      ...........................................................g 168
U07800    237      ............................................................ 178

                                  F    LCRed640
PCR Prod  61       acccccagttcactcaaggagcccggccggatgaacacccataagcatgcccgattgac 120
M22031    72       ............................................................ 131
X58488    167      ............................................................ 108
U07800    177      c........................................................... 118

                                                      < Primer (antisense)
                                                    gcgggtagttcaaacacacggactt
PCR Prod 121      cttcctacgtcgactcgaaatggtccagcaattgat cgcccatcaagtttgtgtgcctgaa 181
M22031    132     ............................................................. 192
X58488    107     ............................................................. 49
U07800    117     ............................................................. 59
```

Fig. 1. Alignment of the amplified fragment within the *B. pertussis* insertion sequence *IS481* (GenBank Accession #M22031, X58488, and U07800). Primer annealing sites, orientation and the sequence of *IS481*-specific hybridization probes are indicated by bold and underlined letters

```
                      Primer >
PCR Prod  1        caccgcctacgagttggagatccaggcccacagcccacaggcggagatcgtctatgactt 60
X66858    733      ............................................................ 792

PCR Prod  61       gttccatgtcgtggccaagtatggacgagaggtcattgatcgggtgcgcgtggatcaggc 120
X66858    793      ............................................................ 852

PCR Prod  121      caatcaactacgccaggatcgtcccgcacgcaggatcatcaaatcgagtcgctggctgct 180
X66858    853      ............................................................ 912

PCR Prod  181      gctgcgcaaccgtgacaacctggatcggcagcaggccgtccggctcgacgaattgctgca 240
X66858    913      ............................................................ 972

PCR Prod  241      agccaaccagccgctgctgacggtctatgtcctgcgtgacgaactcaaacggctctg gtt 300
X66858    973      ............................................................ 1032

                                      F  LCRed705
PCR Prod  301      ctaccaaagacctgcctgggcaagacaagcctggaaccactggtacgagcaggccgagca 360
X66858    1033     ............................................................ 1092

PCR Prod  361      aagcggaatagccgccttgaacaccttcgctcagcgcttgaaaggctatctgcacggcat 420
X66858    1093     ............................................................ 1152

                                          < Primer (antisense)
                                         acttgtggtcgtaacagctcc
PCR Prod   421      cctggccagatgccgacatcccc tgaacaccagcattgtcgagg 464
X66858    1153      ............................................ 1196
```

Fig. 2. Alignment of the amplified fragment within the *B. parapertussis* insertion sequence *IS1001* (GenBank Accession #X66858). Primer annealing sites, orientation and the sequence of *IS1001*-specific hybridization probes are indicated by bold and underlined letters

To complete the amplification mixtures, 18 μl of master mix and 2 μl of the corresponding template DNA preparation were added to each capillary. After a short centrifugation, the sealed capillaries were placed into the LightCycler rotor.

The inclusion of negative as well as positive controls in each set of experiments is considered to be obligatory in the field of diagnostic PCR. The negative control sample was prepared by replacing the DNA template with PCR-grade water. The positive control sample was prepared by adding 2 µl of a mixture of *B. pertussis* and *B. parapertussis* genomic DNA (approximately 1 ng each) to the master mix.

The following PCR protocol was used for amplification and hybridization probe-based detection of the of the *IS481*- and *IS1001*-specific amplicons:

- Denaturation for 10 min at 95°C (to activate the Fast Start *Taq* DNA polymerase)
- Amplification

Parameter	Value		
Cycles	50 (or 40)		
Type	None		
	Segment 1	Segment 2	Segment 3
Target temperature [°C]	95	50	72
Incubation time [s]	10	10	20
Temperature transition rate [°C/s]	20	20	20
Acquisition mode	None	Single	None
Gains	F1=1; F2=15; F3=30		

- Cooling for 2 min at 40°C

As with every LightCycler dual color experiment, readout of LCRed640 values was performed in channel F2, and readout of LCRed705 values was performed in channel F3.

Results

Compared to well-evaluated in-house PCR protocols for the separate detection of *B. pertussis* and *B. parapertussis* in nasopharyngeal samples (amplification in traditional thermocycler devices and product detection in ethidium bromide-stained agarose gels, as well as TaqMan PCR assays), the LightCycler system was examined to determine if it could simplify the diagnostic laboratory workflow by multiplexing, automation, reducing the assay's turn-around time, and reducing the possibility of product contamination frequently associated with post-PCR amplicon manipulation. As with every modified or new PCR protocol in diagnostic microbiology, the sensitivity and specificity of any assay has to be carefully examined before introducing it into routine testing.

Detection Limit of the PCR Assay: Analytical Sensitivity

To determine the assay's lower limit of detection, PCR experiments were performed under duplex conditions (amplification mixture contained four primer and four hybridization probe oligonucleotides) on serial dilutions of genomic DNA prepared from cultured *B. pertussis* and *B. parapertussis* organisms. Looking at the *IS481*-specific sets of primers and hybridization probes, a detection limit of 1 pg of *B. pertussis* template DNA was observed (Fig. 3). For *IS1001*-specific

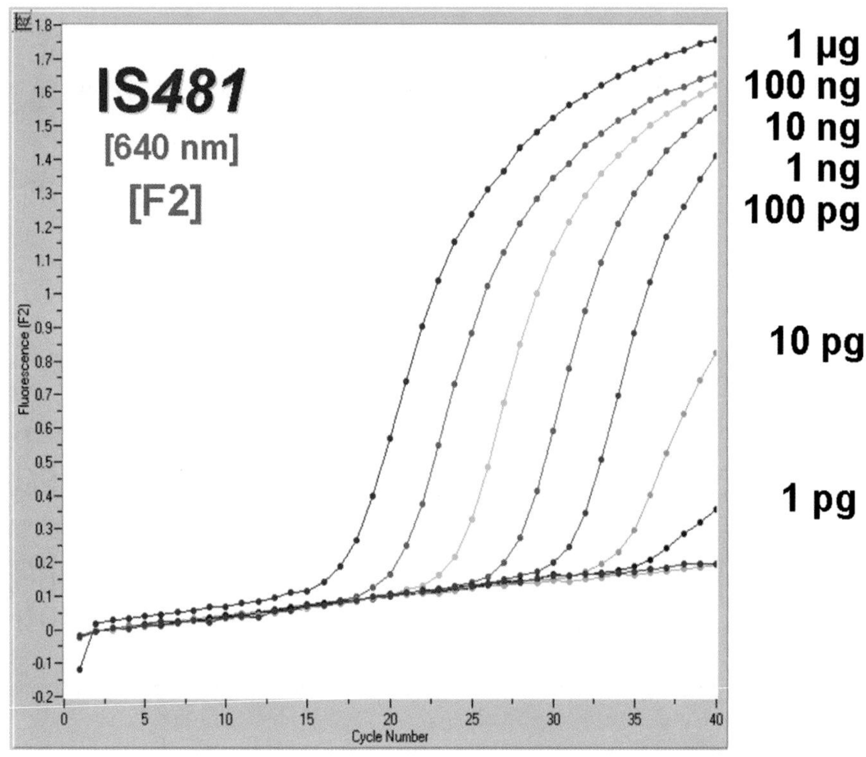

Fig. 3. Sensitivity of the multiplex PCR assay for the detection of *IS481* determined with serial dilutions of *B. pertussis* genomic DNA. For each dilution, the corresponding amount of *B. pertussis* genomic DNA present in the reaction mixture is given next to the amplicon curve

primers and hybridization probes, a detection limit of 1 pg of *B. parapertussis* template DNA was determined (Fig. 4). The diagnostic sensitivity of the assay was further examined with a dilution series performed on a typical nasopharyngeal swab specimen obtained from a *B. pertussis*-positive patient. After 40 cycles of amplification, positive PCR results were observed even with template DNA prepared from a 1:100 dilution of the specimen (data not shown).

Analytical Specificity

The duplex PCR assay was evaluated with a collection of 208 well-characterized clinical samples (see Table 2) originating from different patients presenting with pertussis-like symptoms and a clinical diagnosis of definite or probable pertussis. Previous PCR testing of this collection using different in-house assays has revealed 48 samples positive for *B. pertussis* and 18 samples positive for *B. parapertussis* DNA. Two clinical samples were found positive for *B. pertussis* and *B. parapertussis*. In addition, 24 cultured strains of *B. pertussis* (n=14), *B. parapertussis* (n=5), *B. holmesii* (n=1), *B. trematum* (n=1), and *B. bronchiseptica* (n=3), as well as 80 isolates of Gram-negative and Gram-positive bacterial species other than *Bordetella*

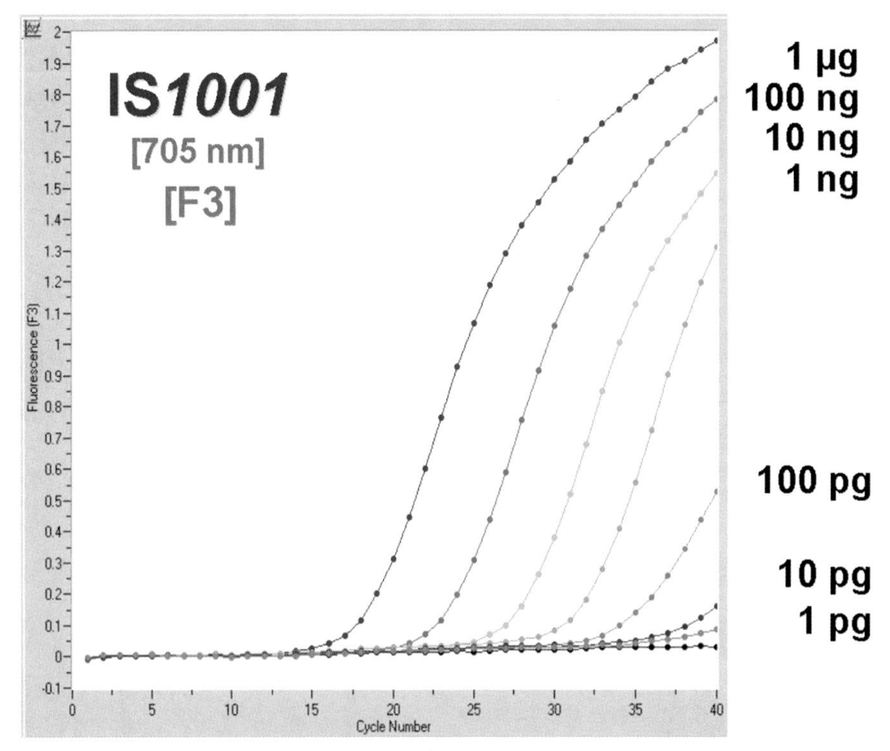

Fig. 4. Sensitivity of the multiplex PCR assay for the detection of *IS1001* determined with serial dilutions of *B. parapertussis* genomic DNA. For each dilution, the corresponding amount of *B. parapertussis* genomic DNA present in the reaction mixture is given next to the amplicon curve

Table 2. Distribution of results by specimen type

Specimen type	Total no. (n=208)[a] of specimens	No. of specimens PCR-positive	No. of specimens Culture-positive
Nasopharyngeal swab	77	17	8
Nasopharyngeal aspirate	76	28	15
Nasal swab	33	8	3
Throat swab	17	5	2
Sputum	5	0	0

[a] Total number of patients, 208.

were examined. The clinical strains of *B. pertussis* and *B. parapertussis* were chosen from different geographical locations and years of isolation to minimize clonality.

LightCycler PCR results with the 208 clinical specimens were in perfect agreement with the results of previous PCR testing and all DNA preparations from cultured *B. pertussis*, *B. holmesii*, and *B. parapertussis* strains showed amplification by increase of fluorescence, whereas none of the DNA preparations from the other cultured bacterial species examined were recognized. Visual examinations of the plots generated by the LightCycler software (cycle number vs fluorescence values of individual capillaries) for channel F2 (*IS481*-specifc amplicons) and channel F3 (*IS1001*-specifc amplicons) provided a clear discrimination between positive and negative samples. The unexpected event of detecting *B. holmesii* by the IS*481*-specific PCR assay is discussed below ("Comments" section). Experimental results on cultured bacteria and a representative set of clinical specimens are shown in Figs. 5 and 6.

Diagnostic Sensitivity and Specificity

As further discussed in the "Comments" section, the lack of a gold standard for detection of *B. pertussis* and *B. parapertussis* in clinical specimens prevented us from calculating reasonable diagnostic specificity and sensitivity values for our assay. However, based on the results of the diagnostic culture and LightCycler PCR with 208 clinical samples, only 45% of the *B. pertussis* PCR-positive specimens were found positive by culture and PCR missed none of the culture-positive specimens. For *B. parapertussis*, only 55% of the PCR-positive specimens were found positive by culture and, again, PCR missed none of the culture-positive specimens (see Table 3). These findings are in agreement with previous studies demonstrating the diagnostic value of PCR for pertussis testing.

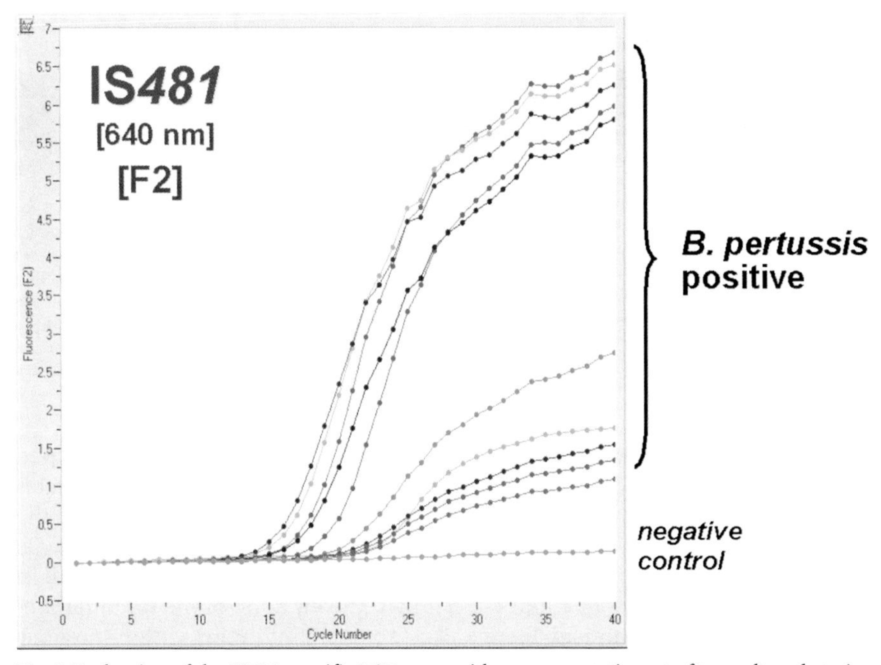

Fig. 5. Evaluation of the *IS481*-specific PCR assay with a representative set of ten cultured strains of *B. pertussis* (including *B. pertussis* ATCC 11615 type strain and nine different patient isolates)

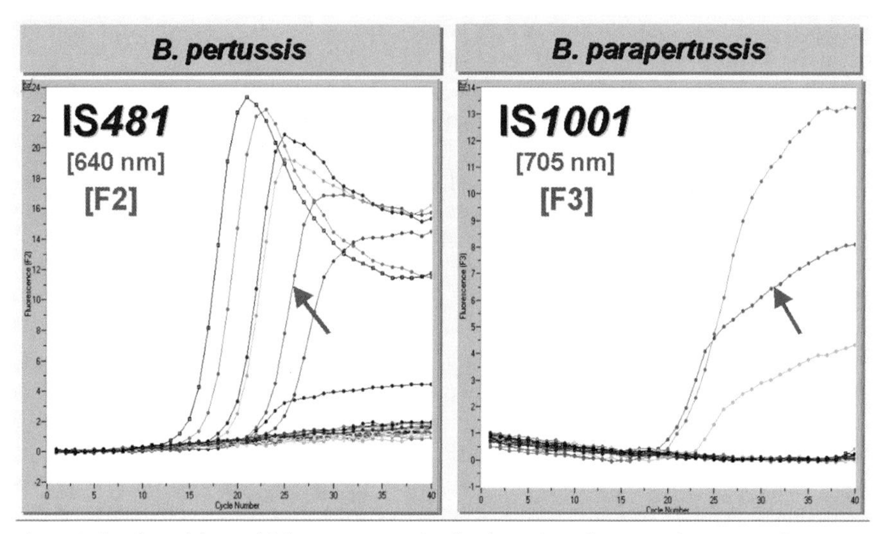

Fig. 6. Evaluation of the multiplex PCR assay for the detection of *IS481* and *IS1001* with a representative set of 32 nasopharyngeal specimens. The clinical specimens originated from different patients presenting with pertussis-like symptoms and a clinical diagnosis of definite or probable pertussis. One patient (curve indicated by a *red arrow*) had a simultaneous infection with *B. pertussis* and *B. parapertussis*

Table 3. Results of *IS481*- and *IS1001*-specific PCR versus results of diagnostic culture

Culture result	No. of patients (n=208)			
	B. pertussis PCR		*B. parapertussis* PCR	
	Positive	Negative	Positive	Negative
Positive[a]	20	0	10	0
Negative[b]	24	164	8	190

[a] Includes two patients whose samples were both positive by PCR and culture for *B. pertussis* and *B. parapertussis.*

[b] Includes two patients whose samples were both positive by PCR and negative by culture for *B. pertussis* and *B. parapertussis.*

Inhibition Events

To investigate PCR inhibition, a separate set of PCR experiments was performed on DNA preparations from 25 respiratory specimens and 5 cultured isolates (spiked with 300 pg of human genomic DNA). Here the primers and probes of the LightCycler Control Kit were used for amplification of the human β-globin gene. No inhibition events were observed with any of the DNA preparations tested. In contrast to direct testing of clinical specimens known to be prone to inhibition, such as blood or stool samples, there was no apparent inhibition with respiratory specimens or cultured bacteria processed with the High Pure PCR Template Preparation Kit.

Comments

LightCycler PCR

Previous methods for diagnostic PCR have employed a variety of time-consuming and laborious procedures (such as Southern blot, DNA sequencing, or solid-phase capturing) for sequence-specific characterization of the amplicons generated with clinical specimens. Enhancing the reliability of the results by sequence-specific probes and simplifying the PCR workflow by a completely automated amplification and online detection procedure with a duplex option, the LightCycler system proved itself as a valuable tool for rapid, sensitive, and simultaneous identification of *B. pertussis* and *B. parapertussis* in the environment of a routine microbiological laboratory. Once DNA is extracted from suitable specimens (such as nasopharyngeal swabs, nasopharyngeal aspirates, perinasal swabs, or throat swabs) and reaction mixtures are completed, PCR results are available within 60 min. The current limitations of the LightCycler system are a total reaction volume of 20 µl (which limits the input of template DNA) and individual sample handling of the reaction cuvettes. The latter limitation will be compensated by using the MagNA Pure LC instrument for DNA preparation and completing PCR reaction mixtures.

Comparison of PCR with Direct Fluorescent-Antibody Testing and Culture

Previous studies suggest that PCR-based assays are comparable or, in certain cases, even superior to culture with respect to sensitivity and specificity [8–12, 15]. Recently the performance of culture, direct fluorescent-antibody (DFA), and PCR testing were compared on 319 consecutive paired specimens [16]. Only 14% of the positive PCR results were confirmed by culture. However, after comparison of the performance of culture, DFA, and PCR against an expanded gold standard defining infection (multiple positive tests or a single positive test other than culture combined with symptoms meeting the CDC clinical case definition [17]), it became apparent that the difference in the PCR and culture positivity rates was due largely to false-negative culture results rather than to false-positive PCR results. Indeed, in this study, the resolved sensitivities of culture and PCR were determined to be 15.2% and 93.5%, respectively. The sensitivity of the DFA was 52.2%.

Applying semiquantitative PCR, another study demonstrated that subjects with a 3+ PCR more frequently experienced typical illness compared with patients with 1+ or 2+ PCR [18]. Here it is worth noting that semiquantitative results will be obtained with the presented LightCycler protocol.

Provided that the method is executed and applied properly, the data from these studies combined with our findings support the suitability of PCR becoming the primary diagnostic test for direct detection of *B. pertussis* and *B. parapertussis* in clinical specimens.

Cross-Reaction of the *IS481*-Specific PCR Assay with *B. holmesii*

In the course of our specificity studies, positive *IS481* PCR results were not only observed with *B. pertussis* isolates but also with the *B. holmesii* ATCC 51541 type strain. This observation is in agreement with recent publications [2, 19, 20], reporting significant cross-reactivity of *IS481*-based PCR assays with *B. holmesii* genomic DNA. By sequencing the corresponding 181-bp amplicons obtained with

B. pertussis and *B. holmesii*, we have recently clarified the molecular reason for this cross-reactivity: both species share identical *IS481* sequences, at least within the amplified region. Due to the presence of *IS481* or *IS481*-like insertion sequences in *B. holmesii*, a simple modification of the assay parameters or switching to other target regions within the 1053-bp *IS481* sequence will not restore specificity for *B. pertussis*. Although the use of alternative targets for *B. pertussis* PCR, such as the pertussis toxin promoter region [18] or genes encoding structural proteins, may enhance assay specificity, these target genes are present singly or in low-copy numbers and may exhibit comparatively lower analytical sensitivity. The potential impact of decreased sensitivity on assay predictive values for pertussis in patients with cough illnesses is uncertain.

From the clinical point of view, however, the risk of *B. holmesii* giving false-positive results by *B. pertussis IS481* PCR should be minimal because the organism has been associated most often with septicemia in patients with underlying conditions [21–24] and up to now it was only found sporadically in patients with respiratory symptoms [19]. *B. holmesii* accounted for less than 4% (32/868) of the *Bordetella* spp. cultured from patients with pertussis-like illness over a 4-year study period [25]. Since the clinical importance and frequency of *B. holmesii* infections has not yet been clearly investigated, the role of *B. holmesii* as a respiratory pathogen remains to be determined [26]. Therefore, *IS481*-based PCR assays could be considered as a valuable addition to pertussis diagnostic testing, especially when used in conjunction with culture and with subsequent species confirmation by phenotypic and biochemical characteristics.

References

1. Heininger U, Stehr K, Cherry JD (1992) Serious pertussis overlooked in infants. Eur J Pediatr 151:342–343
2. Lind-Brandberg L, Welinder-Olsson C, Lagergard T, Taranger J, Trollfors B, Zackrisson G (1998) Evaluation of PCR for diagnosis of *Bordetella pertussis* and *Bordetella parapertussis* infections. J Clin Microbiol 36:679–683
3. Farell DJ, Daggard G, Mukkur TKS (1999) Nested duplex PCR to detect *Bordetella pertussis* and *Bordetella parapertussis* and its application in diagnosis of pertussis in nonmetropolitan Southeast Queensland, Australia. J Clin. Microbiol 37:606–610
4. Deville JG, Cherry JD, Christenson PD, Pineda E, Leach CT, Kuhls TL, Viker S (1995) Frequency of unrecognized *Bordetella pertussis* infection in adults. Clin Infect Dis 21:639–642
5. Centers for Disease Control and Prevention (1993) Resurgence of pertussis: United States 1993. Morbid Mortal Weekly Rep 42:952–960
6. Baron S, Njamkepo E, Grimprel E, Begue P, Desenclos JC, Drucker J, Guiso N (1998) Epidemiology of pertussis in French hospitals in 1993 and 1994: thirty years after a routine use of vaccination. Pediatr Infect Dis J 17:412–418
7. Wirsing von Koenig CH, Rott H, Bogaerts H, Schmitt HJ (1998) A serologic study of organisms possibly associated with pertussis-like coughing. Pediatr Infect Dis J 17:645–649
8. Wadowsky RM, Michaels RH, Libert T, Kingsley LA, Ehrlich GD (1996) Multiplex PCR-based assay for detection of *Bordetella pertussis* in nasopharyngeal specimens. J Clin Microbiol 34:2645–2649
9. Glare EM, Paton JC, Premier R, Lawrence AJ, Nisbet LT (1990) Analysis of a repetitive DNA sequence from *Bordetella pertussis* and its application to the diagnosis of pertussis using the polymerase chain reaction. J Clin Microbiol 28:1982–1987

10. Backman A, Johansson B, Olsen P (1994) Nested PCR optimized for detection of *Bordetella pertussis* in clinical nasopharyngeal samples. J Clin Microbiol 32:2544–2548
11. Schlapfer G, Cherry JD, Heininger U, Uberall M, Schmitt-Grohe S, Laussucq S, Just M, Sterhr K (1995) Polymerase chain reaction identification of *Bordetella pertussis* infection in vaccinees and family members in a pertussis vaccine efficacy trial in Germany. Pediatr Infect Dis J 14:209–214
12. Van der Zee A, Agterberg C, Peeters M, Schellekens J, Mooi FR (1993) Polymerase chain reaction assay for pertussis: simultaneous detection and discrimination of *Bordetella pertussis* and *Bordetella parapertussis*. J Clin Microbiol 31:2134–2140
13. Nygren M, Reizenstein E, Ronaghi M, Lundeberg J (2000) Polymorphism in the pertussis toxin promoter region affecting the DNA-based diagnosis of *Bordetella* infection. J Clin Microbiol 38:55–60.
14. Reischl U, Burggraf S, Leppmeier B, Linde HJ, Lehn, N (2000) Rapid and specific detection of Bordetella pertussis in clinical specimens by LightCycler PCR. In: Meuer S, Wittwer C, Nakagawara K (eds) Rapid Cycle Real-Time PCR: Methods and Applications, Springer Verlag Berlin Heidelberg New York, pp 313–321
15. Tilley PA, Kanchana MV, Knight I, Blondeau J, Antonishyn N, Deneer H (2000) Detection of *Bordetella pertussis* in a clinical laboratory by culture, polymerase chain reaction, and direct fluorescent antibody staining; accuracy and cost. Diagn Microbiol Infect Dis 37:17–23
16. Loeffelholz MJ, Thompson CJ, Long KS, Gilchrist MJR (1999) Comparison of PCR, culture, and direct fluorescent-antibody testing for detection of *Bordetella pertussis*. J Clin Microbiol 37:2872–2876
17. Centers for Disease Control and Prevention (1997) Case definitions for infectious conditions under public health surveillance. Morbid Mortal Weekly Rep 46:25
18. Heininger U, Schmidt-Schlapfer G, Cherry JD, Stehr K (2000) Clinical validation of a polymerase chain reaction assay for the diagnosis of pertussis by comparison with serology, culture, and symptoms during a large pertussis vaccine efficacy trial. Pediatrics 105:E31
19. Yih WK, Silva EA, Ida J, Harrington N, Lett SM, George H (1999) *Bordetella holmesii*-like organisms isolated from Massachusetts patients with pertussis-like symptoms. Emerg Infect Dis 3:441–443
20. Loeffelholz MJ, Thompson CJ, Long KS, Gilchrist MJ (2000) Detection of *Bordetella holmesii* using *Bordetella pertussis* IS*481* PCR assay. J Clin Microbiol 38:467
21. Tang YW, Hopkins MK, Kolbert CP, Hartley PA, Severance PJ, Persing DH (1998) *Bordetella holmesii*-like organisms associated with septicemia, endocarditis, and respiratory failure. Clin Infect Dis 26:389–392
22. Lindquist SW, Weber DJ, Mangum ME, Hollis DG, Jordan J (1995) *Bordetella holmesii* sepsis in an asplenic adolescent. Pediatr Infect Dis J 14:813–815
23. Weyant RS, Hollis DG, Weaver RE, Amin MF, Steigerwalt AG, O'Connor SP, Whitney AM, Daneshvar MI, Moss CW, Brenner DJ (1995) *Bordetella holmesii* sp. nov., a new gram-negative species associated with septicemia. J Clin Microbiol 33:1–7
24. Morris JT, Myers M (1998) Bacteremia due to *Bordetella holmesii*. Clin Infect Dis 27:912–913
25. Mazengia E, Silva EA, Peppe JA, Timperi R, George H (2000) Recovery of *Bordetella holmesii* from patients with pertussis-like symptoms: use of pulsed-field gel electrophoresis to characterize circulating strains. J Clin Microbiol 38:2330–2333
26. Reischl U, Lehn N, Sanden GN, Loeffelholz MJ (2001) Real-time PCR assay targeting IS*481* of *Bordetella pertussis* and molecular basis for detecting *Bordetella holmesii*. J Clin Microbiol 39:1963–1966

Rapid Detection and Simultaneous Differentiation of *Legionella* spp. and *L. pneumophila* in Potable Water Samples and Respiratory Specimens by LightCycler PCR

Nele Wellinghausen*, Olfert Landt, Udo Reischl

Introduction

Legionellae are gram-negative bacteria that are ubiquitous in environmental water sources. They cause epidemic and sporadic cases of pneumonia after inhalation or aspiration of contaminated water droplets. To date, 42 species with 64 serogroups are known [2], many of which cause disease in humans [12, 15, 20]. *Legionella pneumophila* is by far the most common pathogenic species and accounts for up to 90% of the cases of legionellosis [20]. The prognosis of legionellosis is grave. Even with adequate antibacterial chemotherapy, the lethality ranges between 5% and 30%, with a higher case fatality-rate in the elderly and immunocompromised [3].

Infection with *Legionella* spp. accounts for 3%–8% of community-acquired pneumonias [3]. The estimated number of unknown cases may be much higher since the laboratory diagnosis of legionellae is still a challenge. As in most bacterial infections, culture is regarded as the "gold standard" for detection of legionellae. However, due to fastidious growth requirements, overgrowth by other bacteria, and problems inherent with collection and transport of the specimen, the sensitivity of culture is as low as 50%–60% [3]. Furthermore, diagnostic culture for *Legionella* is often requested after initiating antibiotic therapy, which lowers the recovery rate substantially. Serological testing is not helpful in the majority of clinical situations, because even 14 days after the beginning of symptoms seroconversion occurs in only 50% of patients, and 25% of patients with legionellosis fail to develop diagnostic antibody titers [6]. Urinary antigen tests have a sensitivity of 50%–86% but are not capable to detect antigens of all *Legionella* species pathogenic for humans [5, 8].

Since water systems of large buildings, such as hospitals, are often contaminated with legionellae and can thus be the source of nosocomial legionellosis, surveillance of hospital water systems is needed. Isolation of legionellae from water samples by culture techniques is the method usually preferred, but is has the same limitations as described for clinical samples and fails to detect legionellae that reside in amoeba.

* Nele Wellinghausen (✉) (e-mail: nele.wellinghausen@medizin.uni-ulm.de)
Institute of Medical Microbiology and Hygiene, University of Ulm, Robert-Koch-Str. 8, 89081 Ulm, Germany

Recently, PCR techniques for the detection of legionellae in clinical as well as in environmental samples have been developed to overcome the limitations of the conventional diagnostic tests. The most common target genes for the *Legionella*-specific PCR are the 16S rRNA gene, the 5S rRNA gene [7], and the macrophage infectivity potentiator (*mip*) gene [1, 4, 9–11, 13, 14].

Here, we describe the development of a sensitive and specific hybridization probe-based dual-color LightCycler PCR assay for the simultaneous detection of both *Legionella* spp. and *L. pneumophila*. The 16S rRNA gene of *Legionella* was selected as target gene since it exists in multiple copies per genome and contains genus-specific as well as species-specific segments. Two pairs of LightCycler hybridization probes were selected for the sequence-specific detection of amplicons. One pair targets a segment of the 16S rRNA gene conserved among all *Legionella* spp., and the other pair targets a *L. pneumophila*-specific region within the 16S rRNA gene. The dual-color LightCycler assay was validated with a collection of 46 *Legionella* species and serogroups, including all pathogenic species, *Legionella*-like amoebal pathogen (LLAP) 10, and all serogroups of *L. pneumophila*, as well as with 26 isolates of gram-negative and gram-positive bacterial species other than *Legionella*. The PCR assay was further evaluated with 46 hospital water samples and with 26 respiratory specimens of patients with or without *L. pneumophila* infection. A good correlation of PCR results was observed when compared to the results of conventional microbiological testing.

Materials

Equipment

LightCycler instrument (Roche Diagnostics, Mannheim, Germany)
LightCycler software vers. 3.4 (Roche Diagnostics)
Oligo Primer Analysis software (Molecular Biology Insights, Inc., Cascade, CO)

Reagents

Amplification Primers (Interactiva, Ulm, Germany)
Hybridization Probes (TIB MOLBIOL, Berlin, Germany)
QIAamp DNA Mini Kit (Qiagen, Hilden, Germany)

The following reagents were purchased from Roche Diagnostics:
High-Pure PCR Template Purification Kit
LightCycler FastStart DNA Master Hybridization Probes
LightCycler Control Kit DNA

Procedure

Sample Preparation

Starting with cultured bacteria, the QIAamp DNA Mini Kit was used for preparation of chromosomal DNA according to the manufacturer's instructions. Concentration of the eluted DNA was determined photometrically and a 2-μl aliquot of a 10 pg/μl dilution was transferred to PCR.

Potable water samples were collected at three different hospitals. A-100 ml portion of the water samples was centrifuged at 4,000 x g for 30 min, transferred to a sterile 1.5-ml tube and centrifuged at 8,000 x g for 10 min. The pellet was resuspended in 100 µl of the supernatant, centrifuged at 8,000 x g for 10 min, and total DNA was prepared using the QIAamp DNA Mini Kit according to the manufacturer's instructions.

Starting with clinical specimens (bronchoalveolar lavages, DTT-liquefied tracheal or bronchial secretions), the High Pure PCR Template Preparation Kit was used for preparation of total DNA. Briefly, 500-µl aliquots were centrifuged for 2 min at 8000 *x* g, the pellet was suspended in 200 µl of tissue lysis buffer, 40 µl of proteinase K solution (20 mg/ml in dd H_2O) and incubated at 55°C for at least 30 min. After complete disintegration of the specimen, which can be examined visually, 200 µl of binding buffer was added and a further incubation was performed at 70°C for 10 min. Subsequently, 100 µl of isopropanol was added and the mixture was applied to the High Pure spin column. Following the centrifugation and wash steps, total DNA was eluted with 200 µl of elution buffer and a 2-µl aliquot was directly transferred to PCR. The remainder was stored at –20°C for further experiments.

Primer and Hybridization Probe Design

A previously published primer pair [4] (see Table 1) was used for amplification of the *Legionella* spp. 16S rRNA gene. Based on the primer annealing sites, an alignment was performed with different *Legionella* spp. 16S rDNA sequence entries deposited in GenBank and EMBL databases (Fig. 1). The selection of candidate sequences for LightCycler hybridization probes was aided by the Oligo software in order to obtain comparable T_m and GC contents within a range of 35%–60%, and to avoid stable secondary structures, regions of significant self-complementarity and stretches of palindromic sequences. For the detection of *Legionella* spp., a pair of LCRed640-labelled hybridization probes was selected complementary to a region conserved among all *Legionella* spp. For the detection of *L. pneumophila*, a pair of LCRed705-labelled hybridization probes was selected complementary to a *L. pneumophila*-specific region within the ampli-

Table 1. Oligonucleotides

	Length	GC (%)	T_m (°C)
Legionella 16S rRNA gene			
Primers:			
AGGGTTGATAGGTTAAGAGC	20	45.0	49.8
CCAACAGCTAGTTGACATCG	20	50.0	53.4
Probes (*Legionella* spp.):			
AGTGGCGAAGGCGGCTACCT-F	20	65.0	65.1
LCRed640-TACTGACACTGAGGCACGAAAGCGT	25	52.0	64.8
Probes (*Legionella pneumophila*):			
CCAGTATTATCTGACCGTCCCA-F	22	50.0	57.4
LCRed705-TAAGCCCAGGAATTTCACAGATAACTT	27	37.7	62.8

```
                  Primer JRP >
PCR Prod. 1    ccaacagctagttgacatcgtttacagcgtggactaccagggtatctaatcctgtttgct 60
M59157    836  ............................................................ 777
AF129522  614  .........................g.................................. 673
Z49718    795  ............................................................ 736

                   < LegLC (antisense) Red640      F < LegFL (antisense)
                   tgcgaaagcacggagtcacagtcat       tccatcggcggaagcggtga
PCR Prod. 61   ccccacgctttcgtgcctcagtgtcagtattaggccaggtagccgccttcgccactggtg 120
M59157    776  ............................................................ 717
AF129522  674  ............................................................ 733
Z49718    735  ............................................................ 676

PCR Prod. 121  ttccttccgatctctacgcatttcaccgctacaccggaaattccactaccctctcccata 180
M59157    716  ............................................................ 657
AF129522  734  ............................................................ 793
Z49718    675  ............................................................ 616

                             LegPn-HP-1 >     F   Red705    LegPn-HP-2 >
PCR Prod. 181  ctcgagtcaaccagtattatctgaccgtcccaggttaagcccaggaatttcacagataac 240
M59157    656  ............................................................ 597
AF129522  794  .................c........tg.................g.............. 853
Z49718    615  .................c........tg......--.........g.............. 558

PCR Prod. 241  ttaatcaaccacctacgcaccctttacgcccagtaattccgattaacgctcgcaccctcc 300
M59157    596  ..................................................n....n.... 537
AF129522  854  ....................g.............................t......... 913
Z49718    557  ....................g....................................... 498

PCR Prod. 301  gtattaccgcggctgctggcacggagttagccggtgcttcttctgtgggtaacgtccagt 360
M59157    536  .........n.................................................. 477
AF129522  914  ............................................................ 973
Z49718    497  ...........................................................a 438

                     < Primer JFP (antisense)|
                     cgagaattggatagttggga
PCR Prod. 361  taatcagctcttaacctatcaaccct 386
M59157    476  .......................... 451
AF129522  974  ....t..................... 999
Z49718    437  ....t..................... 412
```

Fig. 1. Alignment of the amplified fragment to the *Legionella pneumophila* 16S rRNA gene (GenBank Accession #M59157), the *Legionella dumoffii* 16S rRNA gene (GenBank Accession #AF129522), and the *Legionella bozemanii* 16S rRNA gene (GenBank Accession #Z49718). Primer annealing sites, orientation and the sequence of *Legionella* spp. and *Legionella pneumophila*-specific hybridization probes are indicated by *underlined letters*

con. Annealing sites of primers and hybridization probes within the corresponding GenBank sequences are depicted in Fig. 1. Nucleotide sequences are listed in Table 1.

Inhibition Control

The LightCyler Control Kit DNA was used for detection of *Taq* DNA polymerase inhibitors, possibly present in DNA preparations from clinical samples. The kit provides control DNA template, primers and hybridization probes for amplifying and specific detection of the human β-globin gene. For detection of PCR inhibitors in the water samples, recombinant phage DNA was constructed to serve as an internal inhibitor control. The recombinant molecule consisted of a fragment of lambda DNA flanked by sequences complementary to the primers JFP/JRP [18]. To identify even weak inhibition events, 1 fg of the recombinant lambda control DNA was amplified in each case. The inhibitor control experiments were performed in separate capillaries.

LightCycler PCR

The following master mix was used for amplification and hybridization probe-based dual-color multiplex detection of the *Legionella* 16S rRNA gene:

	Volume [µl]	Final concentration
LightCycler FastStart DNA Master Hybridization Probes	2	1×
$MgCl_2$ stock solution (25 mM)	1.6	3 mM
Primers (10 µM each)	1+1	0.5 µM each
Hybridization probes LegFL/LegLC (2 µM each)	2+2	0.2 µM each
Hybridization probes LegPn-HP-1/LegPn-HP-2 (2 µM each)	2+2	0.2 µM each
H_2O (PCR grade)	4.4	
Total volume	18	

To complete the amplification mixtures, 18 µl of master mix and 2 µl of the corresponding template DNA preparation (water samples: 15 µl of master mix and 5 µl of the DNA preparation) were added in each capillary. After a short centrifugation step, the sealed capillaries were placed into the LightCycler rotor.

In each set of experiments, a negative and a positive control were included. The negative control was prepared by replacing the DNA template with PCR grade water. The positive control sample consisted of 10 pg of *L. pneumophila* SG 1 (ATCC 33152) template DNA.

The following PCR protocol was used for amplification and hybridization probe-based detection of the *Legionella* 16S rRNA gene PCR:

- Denaturation for 10 min at 95°C (to activate the Fast Start *Taq* DNA polymerase)
- Amplification

Parameter	Value		
Cycles	50		
Type	Quantification		
	Segment 1	Segment 2	Segment 3
Target temperature [°C]	95	57	72
Incubation time [s]	0	5	15
Temperature transition rate [°C/s]	20	20	20
Acquisition mode	None	Single	None
Gains	F1=1, F2=15, F3=30		

Readout of LCRed640 values was performed in channel "F2", and readout of LCRed705 values was performed in channel "F3".

The following program was used for the melting curve analysis of the *Legionella*-specific PCR amplicons:

Parameter	Value		
Cycles	1		
Type	Melting curve		
	Segment 1	Segment 2	Segment 3
Target temperature [°C]	95	40	95
Incubation time [s]	0	30	0
Temperature transition rate [°C/s]	20	20	0.1
Acquisition mode	None	None	Continuous

- Cooling for 2 min at 40°C

Results

Analytical Sensitivity

In order to determine the assay's analytical sensitivity, PCR experiments were performed on serial tenfold dilutions of genomic DNA prepared from cultured *L. pneumophila* SG 1 (ATCC 33152). A lower limit of detection of 10 fg of template DNA per PCR assay was determined for both pairs of hybridization probes (Fig. 2A, B).

Specificity

For specificity testing, the dual-color PCR assay was performed with DNA preparations of 46 *Legionella* species and serogroups, including all legionellae pathogenic for humans and LLAP 10. While the *Legionella* genus-specific hybridization probes (LegFL/LC) detected all investigated strains, hybridization probes LegPn-HP-1/-2 specifically discriminated *L. pneumophila* (serogroups 1–15). A clear differentiation between *L. pneumophila* and non-pneumophila legionellae was achieved by melting curve analysis in channel F3. While a characteristic melting temperature (T_m) of 62°C was observed for all *L. pneumophila*-derived amplicons, amplicons from other *Legionella* species were either not recognized, or had T_m values of below 56°C (Fig. 3). Although it has been demonstrated earlier [4] that the PCR assay does not recognize other bacteria typically found in respiratory specimens, a panel of 26 Gram-negative and Gram-positive bacterial strains other than legionellae was investigated. No cross-reactivity was observed in this series of PCR experiments.

Hospital Water Samples

To prove the applicability of the PCR assay on environmental specimens, 46 potable water samples of three different hospitals were investigated by PCR and culture. With the *Legionella* genus-specific probes LegFL/LC, positive PCR results were obtained in 43/46 samples (93.5%), while only 33/46 samples (71.7%) were positive by culture. PCR inhibition events were excluded in all sam-

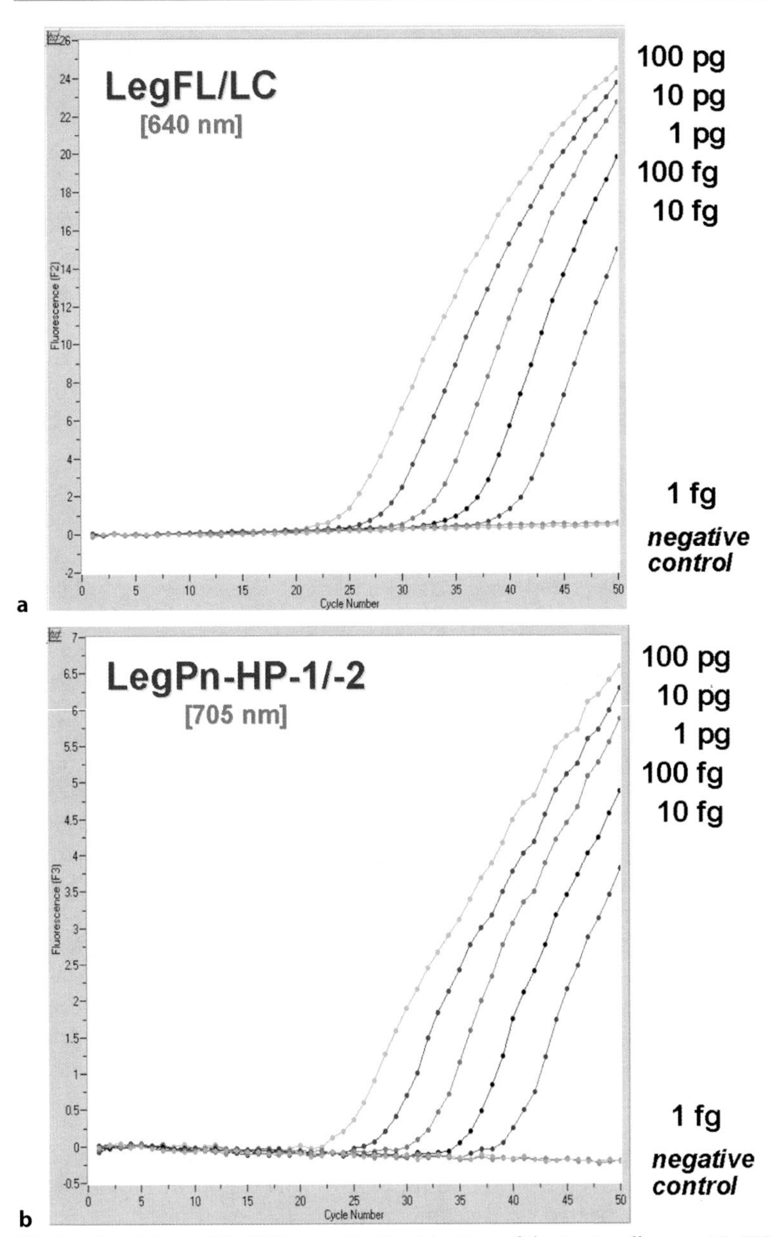

Fig. 2. A Sensitivity of the PCR assay for the detection of the *Legionella* spp. 16S rRNA gene with serial dilutions of *L. pneumophila* SG 1 DNA. For each dilution, the corresponding amount of *L. pneumophila* genomic DNA present in the reaction mixture is given next to the amplicon curve. **B** Sensitivity of the PCR assay for the detection of the *L. pneumophila*-specific 16S rRNA gene sequence with serial dilutions of *L. pneumophila* SG 1 DNA. For each dilution, the corresponding amount of *L. pneumophila* genomic DNA present in the reaction mixture is given next to the amplicon curve

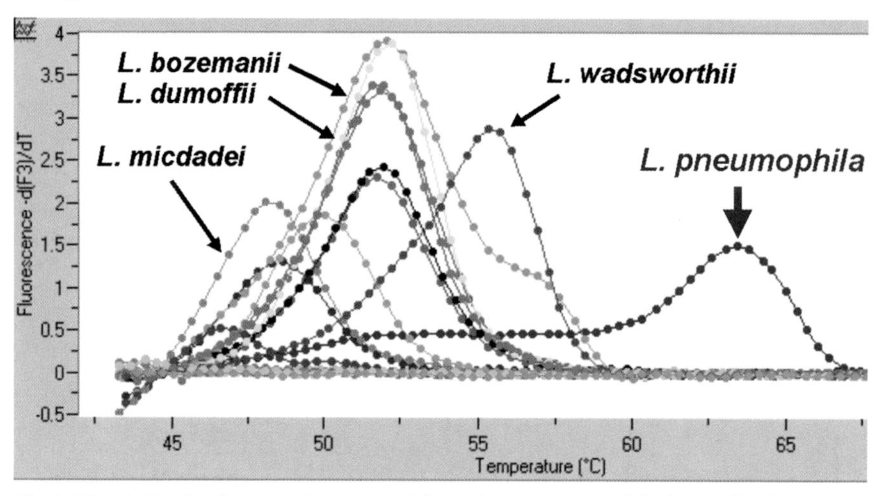

Fig. 3. Discrimination between *L. pneumophila* and non-pneumophila legionellae by melting curve analysis of the PCR product after hybridization with the probes LegPn-HP-1/-2. All known human pathogenic legionellae (20 pg DNA/PCR) were tested. No binding of the probes was observed with *L. jordanis, L. oakridgensis, L. feelei,* and *L. longbeachae*

ples. All but one of the culture-positive samples were also positive by PCR (32/33). Thus, the sensitivity of the PCR for cultural proven *Legionella* contamination was 97.0%. Of the culture-negative samples (13/46), 11 tested positive in PCR suggesting a higher sensitivity of the PCR compared to culture. In 40 of 43 samples that were positive with the *Legionella* genus-specific probes (channel F2), the LightCycler results with hybridization probes LegPnHP-1/-2 (channel F3) revealed the presence of *L. pneumophila*-specific amplicons (T_m at 62°C). Two samples remained negative and one sample revealed the presence of legionellae other than *L. pneumophila* (Fig. 4). Culture was negative for this sample and attempts to sequence the 16S rRNA gene PCR product were not successful. The three LegFL/LC-negative samples were also negative with the *L. pneumophila*-specific probes.

Clinical Specimens

The dual-color PCR assay was further evaluated with a representative set of 26 respiratory specimens originating from 12 different patients presenting with symptoms and a clinical diagnosis of definite or probable legionellosis, and from 14 patients without a suspected *Legionella* infection.

LightCycler PCR results on the clinical specimens were in agreement with the results of traditional microbiological testing (direct fluorescent antibody testing, culture, and/or *L. pneumophila* SG 1 polysaccharide urinary antigen ELISA), including ten specimens positive for *L. pneumophila.* Two BAL samples from patients with clinical symptoms of legionellosis were positive for *L. pneumophila* by PCR testing only. To demonstrate the specificity of melting curve analysis, the

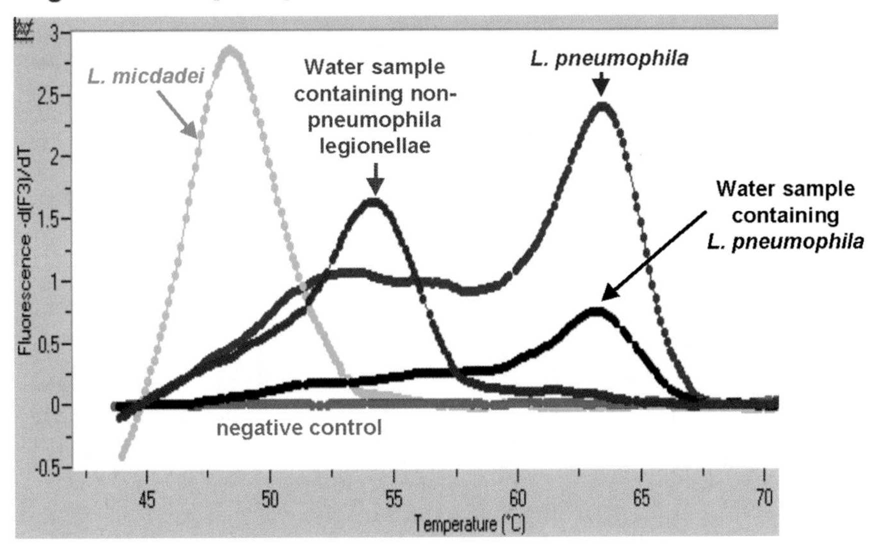

Fig. 4. Detection of *L. pneumophila* and non-pneumophila legionellae in potable water samples. Melting curve analysis of the PCR product after hybridization with the probes LegPn-HP-1/-2 of two representative samples out of 46 with positive and negative controls is shown

experiment included three tracheal secretions (previously negative for *Legionella* spp.) spiked with 1 ng of genomic DNA from *L. longbeachae*, *L. bozemanii*, or *L. micdadei*.

Visual examinations of the plots generated by the LightCycler software (cycle number vs the fluorescence of individual capillaries) allowed for a clear discrimination between positive and negative samples (Fig. 5A). Subsequent melting curve analysis in channel F3 showed a clear discrimination between *L. pneumophila*-derived amplicons (T_m 62°C) and amplicons from non-pneumophila legionellae (T_m below 56°C) (Fig. 5B).

Comments

LightCycler PCR

Previous diagnostic PCR methods have required a variety of time-consuming and laborious procedures, such as Southern blot, DNA sequencing and solid-phase capture for sequence-specific characterization of the PCR products generated from clinical specimens. Increasing test reliability by sequence-specific hybridization probes and simplifying the PCR workflow by automated amplification and real-time detection, the LightCycler system is a valuable tool for the rapid and highly sensitive detection of legionellae in clinical specimens as well as in potable water samples. By the dual-color multiplex option, the simultaneous detection and differentiation between *Legionella* spp. and *L. pneumophila* was possible. As

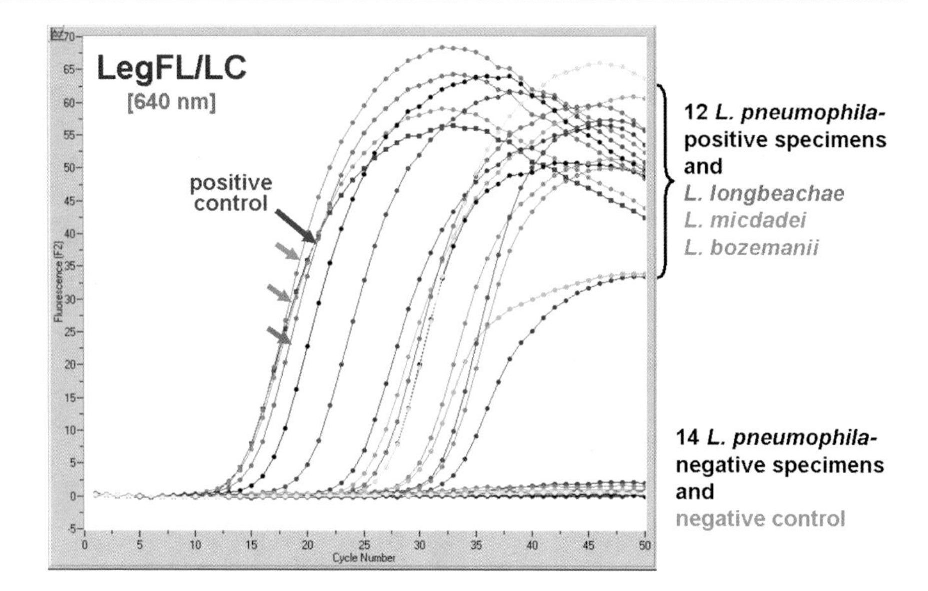

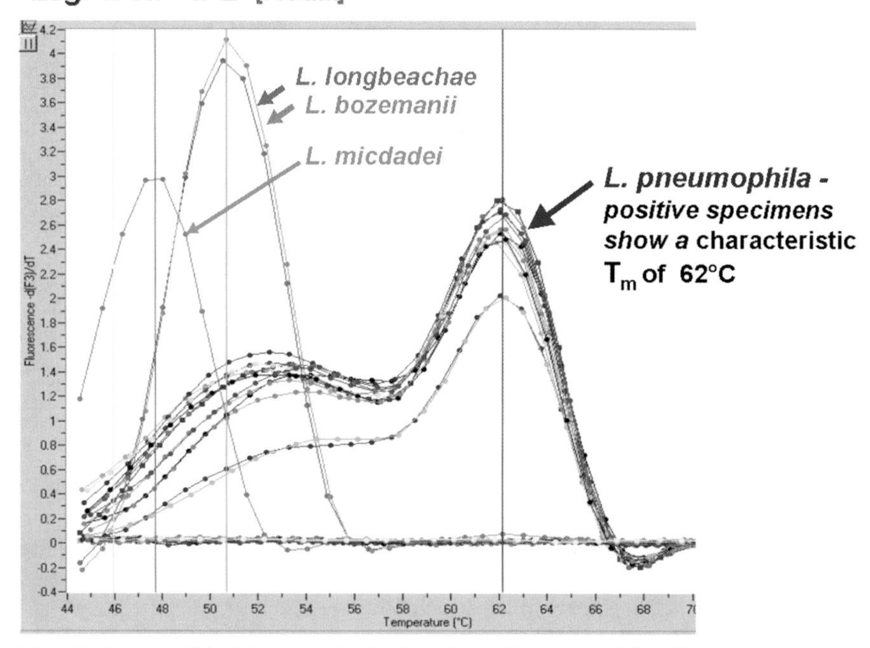

Fig. 5. Evaluation of the PCR assay for the detection of *L. pneumophila* with a representative set of 26 respiratory specimens originating from 12 different patients presenting with symptoms and a clinical diagnosis of definite or probable legionellosis, and from 14 patients without suspected *Legionella* infection (**A**). To demonstrate the specificity of melting curve analysis, 3 tracheal secretions were spiked with 1 ng of *L. longbeachae*, *L. bozemanii*, or *L. micdadei* DNA (**B**)

soon as DNA is extracted from a suitable specimen and reaction mixtures are completed, PCR results are available within less than 50 min. Since the total reaction volume should not exceed 20 µl, the amount of template DNA is limited. However, increasing the input volume of template DNA from 2 µl to 5 µl (per 20 µl total reaction volume) led to a high sensitivity and did not cause PCR inhibition events in the water samples investigated.

Hospital Water Samples

The dual-color multiplex PCR assay for the detection of legionellae and the discrimination of *L. pneumophila* proved to be a sensitive and specific tool for investigation of environmental water samples. Compared to culture results, PCR was more sensitive. This finding is not surprising since recovery rates of culture are usually noticeably less than 100% [17] due to fastidious growth requirements and overgrowth by other bacterial species. In contrast, PCR detection also includes non-culturable and non-viable as well as intra-amoebal legionellae and has already been shown to exceed the sensitivity of culture [11, 16, 19]. The hybridization probes for specific detection of *L. pneumophila* (LegPn-HP-1/-2) did not detect two samples that were positive with the genus-specific probes LegFL/LC, suggesting a lower sensitivity of the *L. pneumophila* probes. However, since these samples were only weakly positive with the probes LegFL/LC and were negative by culture, the degree of legionellae contamination in these samples might not be clinically important. To improve result significance, a quantitative determination of legionellae contamination would be helpful. With the presented protocol, semi-quantitative results can be obtained. Furthermore, the LightCycler instrument and software offer an excellent opportunity to automatically quantify the template DNA by means of an external standard curve with homologous standards. The value of this technique for quantitative investigation of *Legionella* contamination of water samples has already been shown in a recent study [18]. Therefore, further studies together with our findings will prove the significance of LightCycler technology for routine testing of *Legionella* contamination as well as for screening samples during outbreak investigations.

Clinical Samples

Similar to the testing of hospital water samples, our assay proved to be a sensitive and specific tool for investigation of clinical samples. It should be clearly stated that bacterial culture on special growth media (like BCYEα) is considered to be the "gold standard" for diagnosis of *L. pneumophila* infection. Since pulmonary legionellosis may represent a life-threatening disease and the availability of culture results may take up to 2 weeks, direct testing of respiratory specimens by sensitive and rapid LightCycler PCR is an attractive adjunct to the spectrum of conventional diagnostic methods [7].

As with every modified or new PCR protocol in medical microbiology, the diagnostic sensitivity and specificity of the assay has to be carefully examined before introducing it into routine testing. Over the past few years, for example, we have applied a very sensitive *in house* PCR protocol based on chemoluminescent Southern blot detection of *L. pneumophila*-specific amplicons. Due to the high sensitivity of this method, a significant number of "false-positive" results were observed, i.e., a positive PCR result for *L. pneumophila* in the absence of clinical symptoms of the corresponding patient. Laboratory contamination or carry-over events were ruled out in these cases. One possible source of clinical specimen contamination with legionella DNA is tap water which patients use for brushing their teeth. We have detected *L. pneumophila* DNA in the hospital tap water system that is connected to automatic reprocessors used for cleaning and disinfection of bronchoscopes. However, autoclaved and/or ultrafiltered tap water may contain traces of genomic DNA originating from non-viable *L. pneumophila* organisms. These facts should be kept in mind when considering the establishment of a highly sensitive diagnostic PCR assay for *Legionella* spp. Semi-quantitative data, as obtained by LightCycler PCR, should be helpful in assessing the clinical significance of positive results. In the course of our prospective study, we will continue to evaluate the diagnostic value of this LightCycler assay. A detection limit of 10 fg of genomic DNA, which corresponds to about three genome equivalents per PCR reaction, should be sensitive enough for the direct detection of *L. pneumophila* organisms in suitable patient specimens like BAL or tracheal secretions.

References

1. Ballard AL, Fry NK, Chan L, Surman SB, Lee JV, Harrison TG, Towner KJ (2000) Detection of *Legionella pneumophila* using a real-time PCR hybridization assay. J Clin Microbiol 38:4215–4218
2. Benson RF, Fields BS (1998) Classification of the genus Legionella. Semin Respir Infect 13:90–99
3. Breiman RF, Butler JC (1998) Legionnaires' disease: clinical, epidemiological, and public health perspectives. Semin Respir Infect 13:84–89
4. Cloud JL, Carroll KC, Pixton P, Erali M, Hillyard DR (2000) Detection of Legionella species in respiratory specimens using PCR with sequencing confirmation. J Clin Microbiol 38:1709–1712
5. Dominguez JA, Gali N, Pedroso P, Fargas A, Padilla E, Manterola JM, Matas M (1998) Comparison of the Binax Legionella urinary antigen enzyme immunoassay (EIA) with the Biotest Legionella Urin antigen EIA for detection of Legionella antigen in both concentrated and nonconcentrated urine samples. J Clin Microbiol 36:2718–2722
6. Harrison TG, Taylor AG (1988) Timing of seroconversion in Legionnaires' disease. Lancet 2:795
7. Hayden RT, Uhl JR, Qian X, Hopkins MK, Aubry MC, Limper AH, Lloyd RV, Cockerill FR (2001) Direct detection of legionella species from bronchoalveolar lavage and open lung biopsy specimens: comparison of LightCycler PCR, *in situ* hybridization, direct fluorescence antigen detection, and culture. J Clin Microbiol 39:2618–2626
8. Kazandjian D, Chiew R, Gilbert GL (1997) Rapid diagnosis of *Legionella pneumophila* serogroup 1 infection with the Binax enzyme immunoassay urinary antigen test. J Clin Microbiol 35:954–956
9. Kessler HH, Reinthaler FF, Pschaid A, Pierer K, Kleinhappl B, Eber E, Marth E (1993) Rapid detection of Legionella species in bronchoalveolar lavage fluids with the EnviroAmp Legionella PCR amplification and detection kit. J Clin Microbiol 31:3325–3328
10. Koide M, Saito A (1995) Diagnosis of Legionella pneumophila infection by polymerase chain reaction. Clin Infect Dis 21:199–201
11. Koide M, Saito A, Kusano N, Higa F (1993) Detection of *Legionella* spp. in cooling tower water by the polymerase chain reaction method. Appl Environ Microbiol 59:1943–1946
12. Kool JL, Bergmire-Sweat D, Butler JC, Brown EW, Peabody EJ, Massi DS, Carpenter JC, Pruckler JM, Benson RF, Fields BS (1999) Hospital characteristics associated with colonization of

water systems by *Legionella* and risk of nosocomial legionnaires' disease: a cohort study of 15 hospitals. Infect Control Hosp Epidemiol 20:798–805

13. Lindsay DS, Abraham WH, Fallon RJ (1994) Detection of *mip* gene by PCR for diagnosis of Legionnaires' disease. J Clin Microbiol 32:3068–3069
14. Murdoch DR, Walford EJ, Jennings LC, Light GJ, Schousboe MI, Chereshsky AY, Chambers ST, Town GI (1996) Use of the polymerase chain reaction to detect Legionella DNA in urine and serum samples from patients with pneumonia. Clin Infect Dis 23:475–480
15. Neumeister B (1996) Legionella infections: epidemiology, dignostics, clinical aspects, and pathogenesis. Clin Lab 42:715–729
16. Ng DL, Koh BB, Tay L, Heng BH (1997) Comparison of polymerase chain reaction and conventional culture for the detection of legionellae in cooling tower waters in Singapore. Lett Appl Microbiol 24:214–216
17. Villari P, Motti E, Farullo C, Torre I (1998) Comparison of conventional culture and PCR methods for the detection of *Legionella pneumophila* in water. Lett Appl Microbiol 27:106–110
18. Wellinghausen N, Frost C, Marre R (2001) Detection of legionellae in hospital water samples by quantitative real-time LightCycler PCR. Appl Environm Microbiol 67:3985–3993
19. Yamamoto H, Hashimoto Y, Ezaki T (1993) Comparison of detection methods for Legionella species in environmental water by colony isolation, fluorescent antibody staining, and polymerase chain reaction. Microbiol Immunol 37:617–622
20. Yu VL (2000) *Legionella pneumophila* (Legionnaires' Disease), 5th edn. Churchill Livingstone, Pennsylvania

Homogeneous Assays for the Rapid PCR detection of *Burkholderia pseudomallei* 16S rRNA gene on a Real-Time Fluorometric Capillary Thermocycler

Eric P-H Yap*, Soon-Meng Ang, Shirley G-K Seah, Seng-Meng Phang

Introduction

Burkholderia pseudomallei causes a serious infection, melioidosis, in humans, mammals and birds. It is found in the soil and surface water in South East Asia and northern Australia, where human melioidosis is endemic. Sporadic cases of melioidosis have been reported in many regions including Africa, the Indian subcontinent, Europe, Central and South America. The high mortality of this infection is reflected by case fatality rates of 20–60%. Human melioidosis can progress rapidly over hours or persist chronically over decades, and may manifest either as localised skin, hepatosplenic or pulmonary abscesses or as a disseminated septicaemia. It is for these reasons that methods for the early detection of this Gram-negative bacterium are important in clinical diagnosis and management of patients. Laboratory tests for *B. pseudomallei* are also important for epidemiological surveillance, public health, veterinary medicine and environmental studies.

Several PCR methods have been investigated for detecting *B. pseudomallei* 16S rRNA [**1**], 23S rRNA [**2**], 16S-23S intergenic region [**3**] or anonymous sequences. We had previously developed a nested PCR assay for the heat shock protein hsp70 [**4**]. This rapid method was performed on a capillary thermocycler (RapidCycler, Idaho Technology) followed by gel electrophoresis. It allowed detection of as few as 5 CFU's within an hour.

In order to further speed up detection, improve sensitivity and reduce carry-over contamination risk, we explored the use of the capillary fluorometric thermocycler. We report here the development of two rapid homogeneous assays, based on the amplification of specific 16S rRNA sequences, and real-time detection using a generic dsDNA dye or the 5' nuclease assay (known commercially as TaqMan format).

* Eric P-H Yap (✉) (E-mail: nmiv3@nus.edu.sg)
Principal Research Scientist and Research Director
Defence Medical Research Institute, NUS Clinical Research Centre MD 11
10 Medical Drive, Singapore 117597

Materials

Equipment

LightCycler instrument with vers. 3.0 software (Roche Molecular Biochemicals (RMB), Mannheim)
OligoPrimer analysis software (vers. 5.0 for Mac, MBI)
DNAStar DNA analysis software (vers. 2 for Mac, Lasergene)

Kits

LightCycler DNA Master SYBR Green I (RMB)
LightCycler DNA Master Hyb Probes (RMB)
Blood and cell culture DNA extraction kit (Qiagen)

Reagents

Amplification primers (synthesized by Genset, Singapore)
TaqMan probe (synthesized by PE Biosystems)
TaqStart Antibody (Clontech #5400-1/2, Palo Alto, CA)

Cultures

Burkholderia pseudomallei (Reference strain ATCC 23343, clinical and animal isolates from Singapore)
Other bacterial reference and clinical isolates (see Figure 3)

Procedure

Sample preparation

Cultures of *Burkholderia pseudomallei* were grown on LB agar plates overnight at 37C. DNA from individual colonies was extracted using affinity columns (Qiagen) and quantitated photospectrometrically.

Primer and Probe design

B. pseudomallei-specific regions of the 16S rRNA gene were identified using in-house sequence data and public databases. Potential primer sequences were checked for specificity against 16S rRNA from other *Burkholderia spp*, including its closest relative *B. cepacia*, and all entries in Genbank and the Ribosomal Database Project (http://www.cme.msu.edu/RDP/html/index.html). Primers were designed such that there were at least 2 *B. pseudomallei*-specific nucleotides at the 3' end of the primer to confer specificity. Care was taken to minimise primer dimer and hairpin formation, using primer analysis software (Oligo). Two pairs of primers, and a double-labelled 5' nuclease assay probe were designed:

***Burkholderia pseudomallei* 16S rDNA sequence (Genbank Accession #AJ131790)**

	Position	Length	GC (%)	T_m (C)
Primers				
GTCCGGAAAGAAATCAT	489	17	41.2	53.4
CGGTACCGTCATCCACT	538R	17	58.8	60.4
Product	489–538	50		
CATTCTGGCTAATACCCGGAGT	503	22	50.0	66.1
GCCCAACTCTCATCGGGC	1070R	18	66.7	67.4
Product	503–1070	568		
Probe				
Fam- ACCCTGGTAGTCCACGCCCT -Tamra	842	20	65	72.6

LightCycler PCR

SYBR Green I Master mix for each 10µl reaction:

	Volume [l]	[Final]
LightCycler DNA Master SYBR Green I + TaqStart antibody	1.08	1x
$MgCl_2$ (25mM)	0.8	3.0 mM
Primers (10µM each)	0.5+0.5	0.5 M
H_2O (PCR grade)	6.12	–

0.08µl of concentrated TaqStart antibody was added to each 1µl of LightCycler DNA Master SYBR Green I used, and incubated for 5 min in the LightCycler centrifuge adapter rack.

The master mix for each 10µl reaction of the 5' nuclease assay comprised:

	Volume [l]	[Final]
LightCycler DNA Master Hyb probes	1.0	1x
$MgCl_2$ (25mM)	2.0	6.0 mM
TaqMan probe (1µM)	1.0	0.1 M
Primers (10µM each)	0.2+0.2	0.2 M
H_2O (PCR grade)	4.6	

9µl of master mix was added to 1µl of genomic DNA in each capillary.

PCR Protocols

The following PCR protocol was used with SYBR Green I detection:

- **Denaturation**

Denaturation at 95°C for 90 sec.

- **Amplification**

Conventional:

Parameter	Value	
Cycles	35	
Type	Quantification	
	Segment 1	Segment 2
Target temperature[°C]	95	50
Incubation time [s]	0	5
Temperature transition rate [°C/s]	20	20
Acquisition mode	None	Single
Gains	F1=5	

2-stage quick denaturation:

Parameter	Value	
Cycles	5	
Type	Quantification	
	Segment 1	Segment 2
Target temperature[°C]	95	50
Incubation time [s]	0	3
Temperature transition rate [°C/s]	20	20
Acquisition mode	None	Single
Gains	F1=5	

Parameter	Value	
Cycles	30	
Type	Quantification	
	Segment 1	Segment 2
Target temperature[°C]	85	50
Incubation time [s]	0	3
Temperature transition rate [°C/s]	20	20
Acquisition mode	None	Single
Gains	F1=5	

- **Melting curve**

Parameter	Value		
Cycles	1		
Type	Melting Curve		
	Segment 1	Segment 2	Segment 3
Target temperature[°C]	95	50	95
Incubation time [s]	0	30	0
Temperature transition rate [°C/s]	20	20	0.2
Acquisition mode	None	None	Cont.
Gains	F1=5		

For TaqMan detection, the same denaturation protocol was used, followed by an amplification protocol as follows:

Parameter	Value	
Cycles	40	
Type	Quantification	
	Segment 1	Segment 2
Target temperature[°C]	95	60
Incubation time [s]	0	40
Temperature transition rate [°C/s]	20	20
Acquisition mode	None	Single
Gains	F1=25	

No melting curve protocol was used for the TaqMan format.

Melting curve and quantitation analysis was performed according to the LightCycler manufacturer's protocols.

Results

Double stranded DNA-specific dye (SYBR Green I) detection of 50bp amplicon

A PCR assay was developed to a *B. pseudomallei*-specific 50bp segment from the 16S rRNA gene on the capillary thermocycler. The primer annealing temperature and Mg++ concentration was optimized so that there was no unwanted non-specific amplification. *In situ* melting curve analysis of SYBR Green I stained product revealed a single –dF1/dT peak at 79.1°C with intra-experiment variation of 0.1°C **(Fig. 1)**. Primer-dimer formation was minimised by the use of TaqStart antibody to prevent primer extension at substringent temperatures (hotstart PCR). Thus, accumulation of PCR product could be monitored in real-time using the dsDNA-specific dye SYBR Green I **(Fig. 2)**. Without hotstart PCR, primer-dimer formation was significant and melting curve analysis was required to distinguish primer-dimers from the slightly larger specific product (data not shown).

In addition to the conventional thermocycling protocols shown in these figures, a two stage quick denaturation protocol was also used. After an initial 90s denaturation and the first 5–10 cycles of denaturation at 95°C (required for genomic template), a lower denaturation temperature of 85°C (adequate for the short amplicon template) was used for subsequent cycles. This has previously been shown to preserve the enzymatic integrity of *Taq* polymerase, resulting in higher PCR yields and less non-specific amplification from genomic template [5]. These modifications allowed shorter cycle times (35 cycles could be completed in under 14 min) and are suitable for short amplicons on the LightCycler instrument.

The number of cycles was optimised so that as little as 10fg (1.5 genomic equivalents) of purified bacterial DNA could be consistently detected. Below this limit-

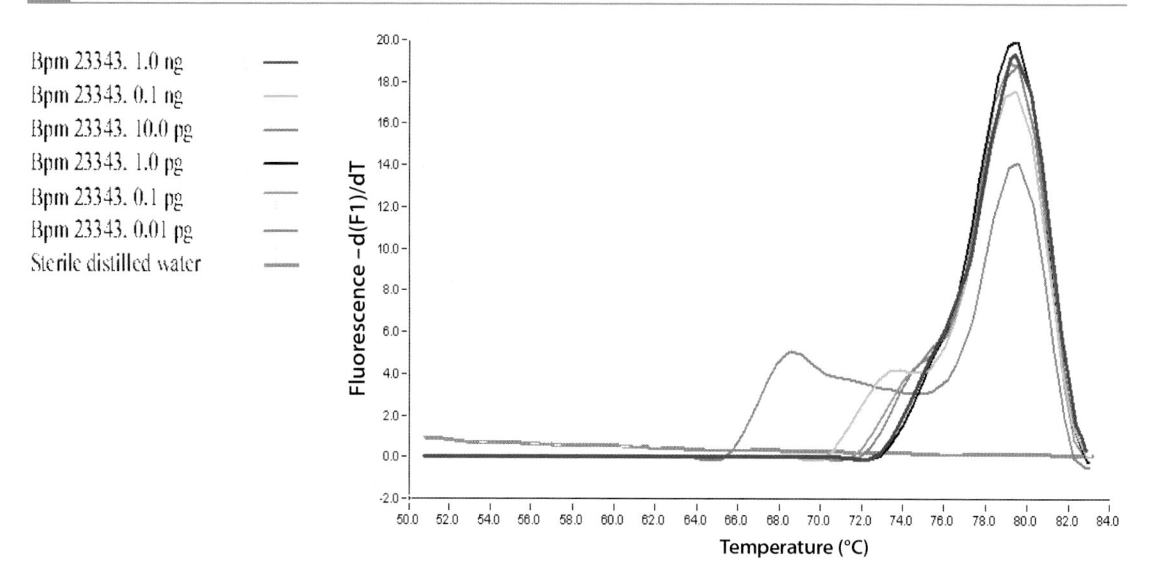

Fig. 1. Melting curve analysis of 10-fold dilutions of different *Burkholderia pseudomallei* concentrations, ranging from 1ng to 0.01pg of genomic DNA, with a specific melting peak at 79.°C (SD = 0.056)

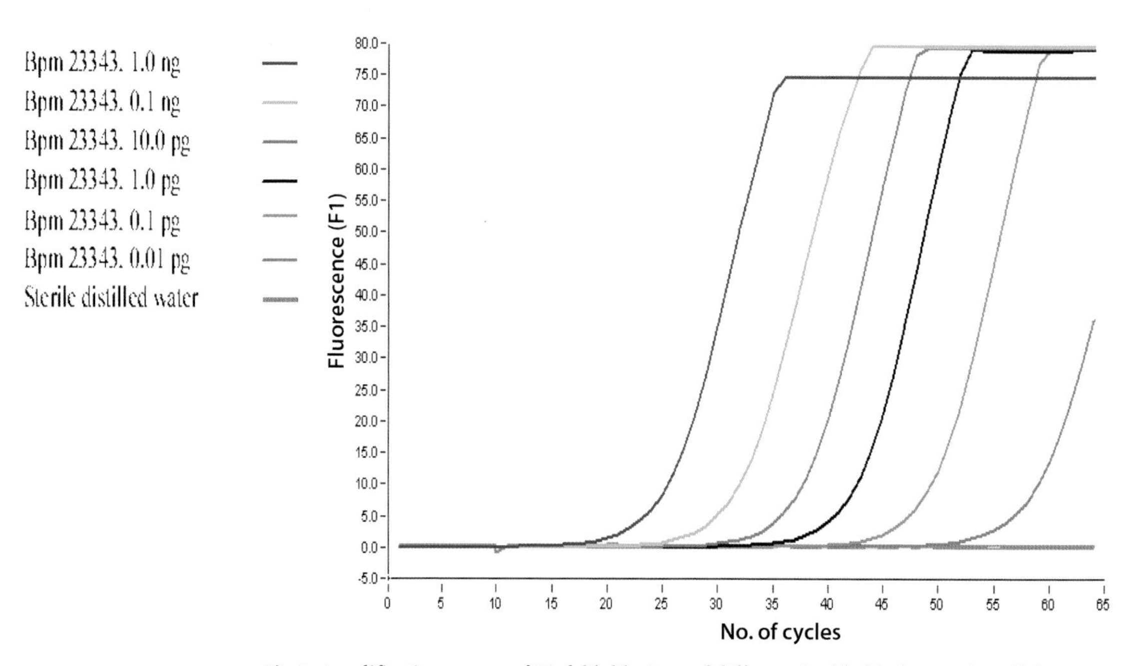

Fig. 2. Amplification curves of 10-fold dilutions of different *Burkholderia pseudomallei* concentrations, ranging from 1ng to 0.01pg of genomic DNA

ing amount, the stochastic effects of dilution result in inconsistent amplification. The dynamic range of the assay covered 7 orders of magnitude, from 100ng (data not shown) through 1ng to 10fg **(Figs. 1, 2)**. The threshold cycle, at which PCR product could first be detected in the F1 channel, correlated tightly with the starting template amounts, and varied between 15 and 45. The expected 50bp PCR product was seen for all samples when run on an agarose gel (data not shown). It was noted that the final yield of product (in fluorescence units) did not reflect initial template amounts.

To evaluate the specificity of the assay, DNA from clinical and reference strains of 23 other bacteria was tested. Despite the use of 100ng (10^7 fold) amounts of these other DNA, no false positives were obtained with these Gram negative and positive bacteria, and with human DNA. No increase from baseline fluorescence was observed, and background fluorescence was established to be due to staining of genomic DNA **(Fig. 3)**. The assay has also been tested on more than 80 human clinical, animal and environmental isolates of *B. pseudomallei*, and all were positive with this assay (data not shown).

5' Nuclease assay (TaqMan probe) detection of 568bp amplicon

A *B. pseudomallei*-specific 568bp PCR product was obtained with a second pair of primers **(Fig. 4)**. To provide a second level of specificity, a generic 16S rRNA probe was used. In its native unbound state, fluorescent emission from the 5' FAM label would be quenched by absorption by the 3' TAMRA fluorophore. During the combined annealing-extension step of each cycle, the hybridized probe would be cleaved by the 5' exonuclease activity of *Taq* polymerase. The accumulation of free FAM could be monitored real-time by fluorescence at 530nm (F1 channel) **(Figs. 4, 5)**. Background emission of TAMRA is negligible at this wavelength. The use of this probe allowed this assay to be highly specific, with no non-specific signal with *B. cepacia* (a close relative of *B. pseudomallei*), *N. gonorrhoea* (also in the β subdivision of Proteobacteria), and human DNA even after over-amplification to 80 cycles **(Fig. 5)**. The detection limit of the 5' nuclease assay was 100fg (15 genome equivalents), with a dynamic range of at least 6 orders of magnitude (100fg-100ng) **(Fig. 4)**. Threshold cycles correlated with starting template amounts, making this assay semi-quantitative. When these PCR products were run on a gel, the expected 566bp PCR product was seen with no contaminating products or primer dimers.

Comments

For molecular diagnostic applications, the possibility of false positive results due to contamination by carryover of products from a previous reaction is a major limitation to the routine use of PCR-based assays **[6]**. The development of homogeneous assays whereby fluorogenic reactions can be monitored directly without any post-PCR manipulation will help to reduce contamination risk. In addition, homogeneous assays offer potential advantages of speed, consistency and less hands-on time. PCR of small amplicons (50bp) is feasible since they do not need to be resolved on gels, further reducing cycling time. The capillary

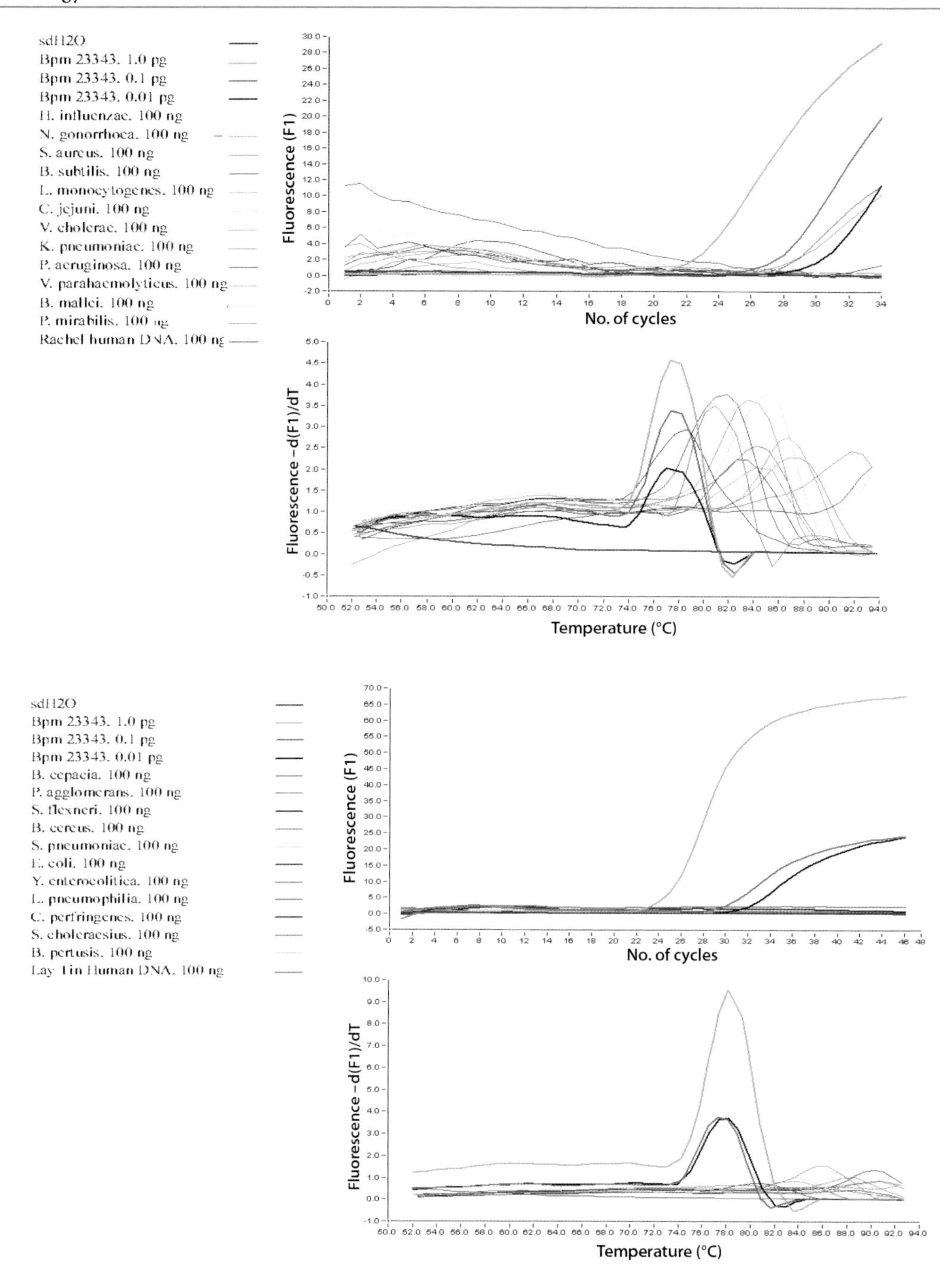

Fig. 3. Specificity test of the 16S rDNA primers for *Burkholderia pseudomallei* with different bacteria and human DNA illustrated by SYBR Green I quantification, and melting curve analysis

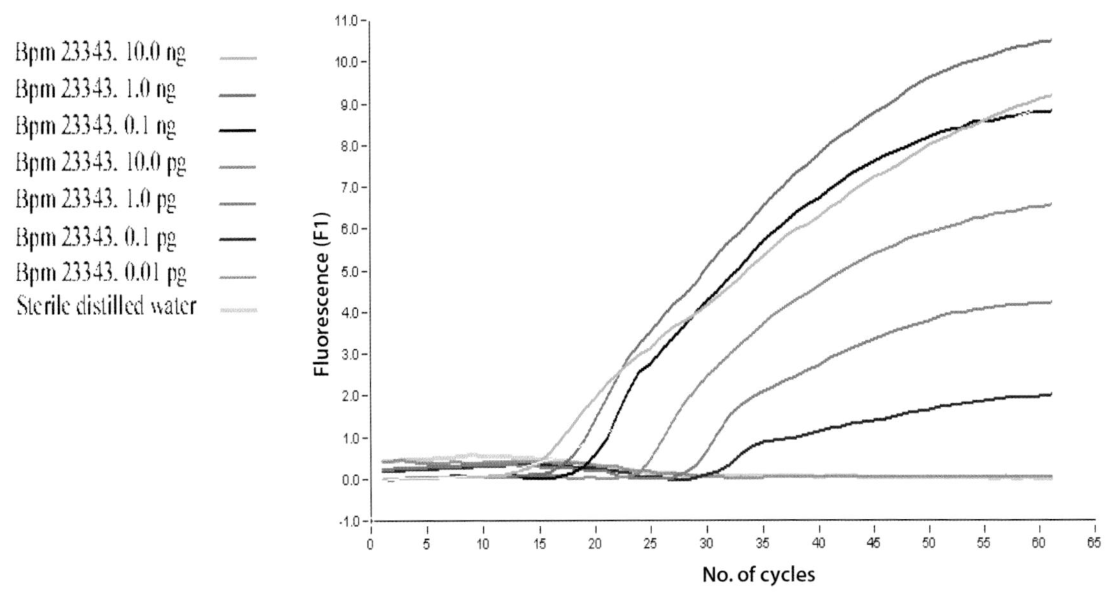

Fig. 4. Quantification curve for amplification utilizing TaqMan detection of PCR products, ranging from 10ng to 0.01pg of *Burkholderia pseudomallei* genomic DNA

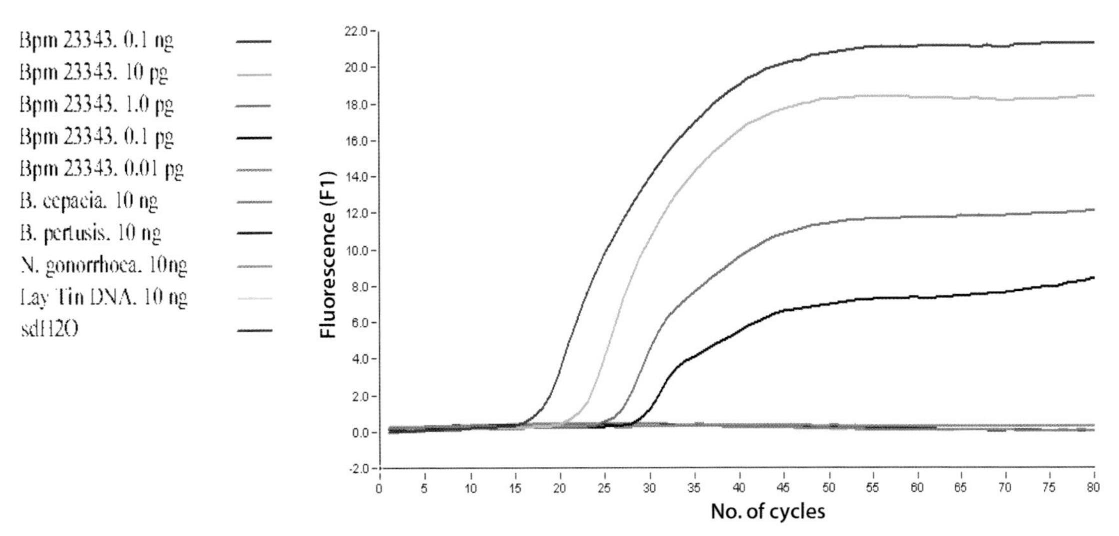

Fig. 5. Specificity testing with the TaqMan detection format with *Burkholderia cepacia, Bordetella pertussis, Neisseria gonorrhoea, and human DNA*

fluorometer-thermocycler (LightCycler) allows homogeneous assays to be performed rapidly, and compared to the earlier generation capillary thermocycler (RapidCycler), has better performance, sensitivity and ease of use.

The *B. pseudomallei* 16S ribosomal RNA gene was selected as an amplification target for several reasons. It is highly conserved between different isolates of *B. pseudomallei* (data not shown). Sequence data from a large number of other bacteria is also available to assist in primer design. It exists in multiple copies in bacterial genomes. Alignment of *B. pseudomallei* with other bacterial 16S sequences allowed highly specific primers to be designed. These primers have been empirically demonstrated to be specific and do not amplify *B. cepacia* which is sometimes mistakenly identified as *B. pseudomallei* by conventional microbiological techniques.

Two assay protocols were developed. The *SYBR Green I* protocol relies on primer specificity and inhibition of primer-dimerization. It does not require labeling of probe (lower cost), has detection sensitivity down to the single cell level (more sensitive), and due to its small amplicon of 50bp is highly efficient and has short cycling times (more rapid). Real-time monitoring reveals an increase of fluorescent signal from as early as 12 cycles (5 min using the quick denaturation protocol), while the complete assay can be completed within 14 min. Melting curve analysis can be used to confirm product specificity. This assay is useful where cost, sensitivity, quantification and rapidity are important concerns.

The *TaqMan* probe protocol involves the amplification of a larger 568bp product (smaller amplicons were desirable but limited by availability of specific primers). This resulted in longer cycling time (~45min for entire assay) and slightly lower sensitivity. However the use of a 16S rRNA probe allowed confirmation of amplicon identity and made the assay resistant to non-specific products and primer-dimers. This assay is useful for large scale screening of samples and for confirmation of positive results from the *SYBR Green I* protocol.

Further improvements to the assays that could be explored include optimizing the components of the master mix to improve PCR efficiency and fluorescence yields, including a built-in positive control, multiplexing several targets, enhancing signal:noise ratios of TaqMan probes, using other probing strategies and adaptation to high throughput microtitre plate format.

In addition to these two assays, other real-time detection formats such as those utilizing hybridization probes based on the FRET (fluorescence resonance energy transfer) and molecular beacons are currently being evaluated for use on the LightCycler instrument. Preliminary results show that these alternative methods are comparable in sensitivity and speed.

The development of homogeneous assays for *B. pseudomallei* may be useful for early diagnosis of primary melioidosis in patients before seroconversion or before culture results are available, for monitoring response to antimicrobial therapy, and for diagnosis of secondary relapse. Their clinical utility needs to be established. These assays may also be useful for animal and environmental studies.

References

1. Dharakul, T, Songsivilai, S, Viriyachitra, S, Luangwedchakarn, V, Tassaneetritap, B and Chaowagul, W (1996) Detection of *Burkholderia pseudomallei* DNA in patients with septcaemic melioidosis. *J. Clin. Microbiol.* 34:609–14
2. Lew, AE and Desmarchelier, PM (1994) Detection of *Pseudomonas pseudomallei* by PCR and hybridisation. *J. Clin. Microbiol.* 32: 1326–32
3. Kunakorn, M and Markham, RB (1995) Clinically practical seminested PCR for *Burkholderia pseudomallei* quantitated by enzyme immunoassay with and without solution hybridization. *J. Clin. Microbiol.* 33:2131–5
4. See, LH, Yap EH and Yap, EPH (1997) Detection of *Burkholderia pseudomallei* by PCR. *Asia Pacific J. Pharmacol.* 12:S27
5. Yap, EPH and McGee, JO'D (1991) Short PCR product yields improved by lower denaturation temperatures. *Nucl. Acids Res.* 19:1713
6. Yap, EPH, Lo, YMD, Fleming, KA and McGee, JO'D False positives and contamination in PCR. In Griffin, HG and Griffin, AM (Eds) *PCR technology: current innovations.* CRC Press, Boca Raton, FL.

Rapid Detection of Toxigenic *Corynebacterium diphtheriae* by LightCycler PCR

Udo Reischl*, Elena Samoilovich, Valentina Kolodkina,
Norbert Lehn, Hans-Jörg Linde

Introduction

Diphtheria is highly contagious and causes localized inflammatory lesions of the upper respiratory tract or skin, with associated necrosis at distant sites attributable to the dissemination of the major virulence factor, the diphtheria toxin. The causative bacterial organism, *Corynebacterium diphtheriae*, was first isolated by Löffler in 1883, and the diphtheria toxin was first described in 1888 by Roux and Yersin. The gene responsible for the production of the 62-kDa exotoxin is carried on a specific bacteriophage. Death can result from respiratory obstruction by pharyngeal membranes or from myocarditis caused by the toxin.

In the past century, diphtheria dramatically decreased in developed countries through the widespread use of active immunization with diphtheria antitoxin. After more than 30 years of excellent control, epidemic diphtheria has reemerged in 1990 in the newly independent states of the former Soviet Union. About 50,000 cases and 1,750 deaths have been reported in 1994 for this geographic region and at least 20 imported cases of diphtheria were observed in the neighbouring and western European countries [1, 2]. Due to a massive immunization campaign, the incidence has decreased in the meantime and the Russian epidemic appears to be under control. While the number of diphtheria cases reported in the United States, Japan, or western Europe has significantly declined, studies of diphtheria immunity levels among adults in the United States have shown that many adults (from 20 to 90%) do not posess adequate immunity against this disease [3]. This could easily result in a resurgence of this fatal disease even in developed countries. Accordingly, the Russian epidemic and waning immunity in western European countries has increased both clinical and microbiological awareness in the diagnosis of diphtheria.

The most prominent difficulties with the laboratory diagnosis of unsuspected diphtheria infection are that colonies of *C. diphtheriae* are not distinctive on culture media commonly used for plating wound or throat cultures and that the presence of coryneform organisms, isolated among other throat or skin flora, is usually not reported. For cases of suspected diphtheria, antitoxin and antibiotics

* Udo Reischl (✉) (e-mail: Udo.Reischl@klinik.uni-regensburg.de)
Institute of Medical Microbiology and Hygiene, University of Regensburg,
Franz-Josef-Strauß-Allee 11, 93053 Regensburg, Germany

are usually given immediately. Then material for culture is obtained on a swab from the inflamed areas of the membranes that are formed in the throat and nasopharynx. For direct demonstration of the organism, smears are made of the original material for subsequent Neisser or Löffler methylene blue stain for observation of metachromatic granules for tentative identification, and Gram stain for coryneform morphology (which is, however, indistinguishable from that of other *Corynebacterium* spp.). The sensitivity of culture may be increased by inoculation of Löffler or Pai slants in addition to Tinsdale or other tellurite-containing media, which is particular neccessary if no liquid enrichment medium was used for transport. For enhanced recovery, specimens submitted as single swabs should be transferred to a serum- or blood-containing medium that can be subcultured after four days of incubation onto the media listed above. Culture of *C. diphtheriae* may also be compromised by the fact that special selection media are not at hand when required, due to the rare neccessity and short shelf life of the media.

The *C. diphtheriae* group (with four cultural variants or subspecies) is distinguished by its ability to produce the diphtheria toxin [4, 5]. During the workup of suspected cultured isolates, at least 10 colonies should be tested for toxigenicity, as both toxigenic and nontoxigenic varieties may be found in the same patient. Starting with a heavily inoculated broth culture of the isolate to be tested, the detection of the toxins has traditionally be performed on an Elek plate on which the bacterial suspension is streaked on an antitoxin-containing filter paper, modifications thereof [6] or, more reliable, in tissue culture cells, where the cytotoxic effect may be seen under the microscope within 24 hours.

The Russian diphtheria epidemic, sporadic cases outside Russia, and a recent report on the misidentification of toxigenic *C. diphtheriae* as a *Corynebacterium* sp. with low virulence in a child with endocarditis [7], has emphasized the need for rapid, simple, and reliable diagnostic methods available for clinical microbiology laboratories. Like in the case of other pathogenic organisms of immense public health significance, PCR has evolved as the method of choice for rapid culture confirmation or, perhaps more important, the direct detection of toxigenic *C. diphtheriae* organisms from clinical specimens.

A number of protocols have been published for the PCR-based detection of the diphtheria toxin gene (*tox*) that codes for the A and/or B subunits of diphtheria toxin [8–13]. Significant genome diversity among *C. diphtheriae* organisms has recently been demonstrated by ribotyping, multilocus enzyme electrophoresis, PCR single-strand conformation polymorphism analysis, and DNA sequencing [14]. Since nucleotide sequence variations within the target gene could have significant effects on the reliability of diagnostic PCR, primers and LightCycler hybridization probes were selected from regions which were shown to be conserved at the sequence level in 79 *C. diphtheriae* strains from Russia and Ukraine [15].

Here we describe the development of a rapid and specific hybridization probe-based LightCycler assay and its validation for the detection of toxigenic *C. diphtheriae* organisms. Testing a collection of 45 respiratory specimens, cultured strains of toxigenic and nontoxigenic *C. diphtheriae*, as well as 80 isolates of Gram-negative and Gram-positive bacterial species other than *C. diphtheriae*, our results were in perfect agreement with the results of previous PCR and conventional microbiological testing.

Materials

Equipment

LightCycler instrument (Roche Diagnostics, Mannheim, Germany)
LightCycler software vers. 3.3 (Roche Diagnostics)
LightCycler Capillaries, Centrifuge Adapters, and Cooling Block (Roche Diagnostics)

Reagents

Oligo Primer analysis software (Molecular Biology Insights, Inc., Cascade, CO)
Amplification Primers (Metabion, Munich, Germany)
Hybridization Probes (TIB MOLBIOL, Berlin, Germany)
The following reagents were purchased from Roche Diagnostics:
High-Pure PCR Template Purification Kit
LightCycler-FastStart DNA Master Hybridization Probes
LightCycler-Control Kit DNA

Procedure

Preparation of Template DNA

Bacterial DNA was prepared, using the High Pure PCR Template Preparation Kit according to the manufacturer's instructions. Starting with nasopharyngeal or throat swabs, about half of the swab tip was cut off, suspended in 200 µl of tissue lysis buffer and 40 µl of proteinase K solution (20 mg/ml) and incubated at 55°C for at least 30 min. After complete disintegration of tissue pieces, which can be examined visually, 200 µl of binding buffer were added and a further incubation was performed at 70°C for 10 min. Subsequently, 100 µl of isopropanol were added and the mixture was applied to the High Pure spin column.

Starting with cultured bacteria, one to 10 colonies were suspended in 200 µl of PBS buffer, 15 µl of a lysozyme solution (10 mg/ml in Tris-HCl, pH 8.0) was added and incubated at 37°C for 10 min. After adding 200 µl of binding buffer and 40 µl of a proteinase K solution (20 mg/ml) with a further incubation at 70°C for 10 min, 100 µl of isopropanol were added and the mixture was applied to the High Pure spin column. Following the centrifugation and wash steps, bacterial DNA was eluted with 200 µl of elution buffer and a 2 µl aliquot was directly transferred to PCR. The remainder was stored at -20 °C for further experiments.

Primer and Hybridization Probe Design

A previously published primer pair [11–13] (*see* Table 1) was used for amplifying a 249-bp segment within the gene encoding for the diphtheria toxin A subunit [16], which proved to be a suitable target to identify toxigenic *C. diphtheriae* organisms. Based on the primer annealing sites, alignments were performed with all of the different *C. diphtheriae* toxin gene sequence entries deposited in GenBank and EMBL databases to recognize and consider single nucleotide ambiguities. The selection of candidate sequences for LightCycler hybridization probes was aided by the Oligo software in order to obtain comparable T_m, GC contents

Table 1. *C. diphtheriae* toxin gene (*tox*) encoding the A subunit of the toxin.

	Length	GC (%)	T_m (°C)
Primers [11]			
GAAAACTTTTCTTCGTACCACGGGACTAA	29	41.4	75.2
ATCCACTTTTAGTGCGAGAACCTTCGTCA	29	44.8	78.6
Probes			
AATAAATACGACGCTGCGGGATAC-F	24	45.8	71.6
LCRed 640-CTGTAGATAATGAAAACCCGCTC	23	43.5	65.4

```
                   Primer >
PCR Prod  1     gaaaacttttcttcgtaccacgggactaaacctggttatgtagattccattcaaaaaggt 60
U11011    46    ............................................................ 105
K01723    418   ............................................................ 477

PCR Prod  61    atacaaaagccaaaatctggtacacaaggaaattatgacgatgattggaaagggttttat 120
U11011    106   ............................................................ 165
K01723    478   ............................................................ 537

                                                F LCRed640
PCR Prod  121   agtaccgacaataaatacgacgctgcgggatactctgtagataatgaaaacccgctctct 180
U11011    166   ............................................................ 225
K01723    538   ............................................c............... 597

                                                  < Primer (antisense)
                                                actgcttccaagagcgtgat
PCR Prod  181   ggaaaagctggaggcgtggtcaaagtgacgtatccaggactgacgaaggttctcgcacta 240
U11011    226   ............................................................ 285
K01723    598   .............a....................................c......... 657

                tttcaccta
PCR Prod  241   aaagtggat 249
U11011    286   ......... 294
K01723    658   ......... 666
```

Fig. 1. Alignment of the amplified fragment within the *C. diphtheriae* toxin gene *(tox)* encoding the A subunit of the toxin (GenBank Accession # U11011, and K01723). Primer annealing sites, orientation and the sequence of *C. diphtheriae* toxin gene-specific hybridization probes are indicated by bold and underlined letters

within the range of 35–60%, and to avoid stable secondary structures, regions of significant self-complementarity, stretches of palindromic sequences and close proximity between the primers and the probes. For the sequence-specific detection of *C. diphtheriae* toxin gene amplicons, a set of hybridization probes was selected (*see* Table 1). Annealing sites of primers and hybridization probes within the corresponding GenBank sequences are depicted in Figure 1.

Inhibition Control

The LightCycler Control Kit DNA was used for the detection of *Taq* DNA polymerase inhibitors, possibly present in DNA preparations from clinical samples. The latter kit provides control DNA template, primers and hybridization probes for amplifying and specific detection of the human ß-globin gene. To identify even weak inhibition events, 300 pg of human genomic DNA was amplified in each case.

LightCycler PCR

The following master mix was used for amplification and hybridization probe-based detection of the *C. diphtheriae* toxin gene-specific amplicons:

	Volume [µl]	[Final]
LightCycler Fast Start DNA Master Hybridization Probes	2	1 x
$MgCl_2$ stock solution (25mM)	1.6	3 mM
Primers (10 µM each)	1+1	0.5 µM each
Hybridization probes (2 µM each)	2+2	0.2 µM each
H_2O (PCR grade)	8.4	
Total volume	18	

To complete the amplification mixtures, 18 µl of master mix and 2 µl of the corresponding template DNA preparation were added in each capillary. After a short centrifugation, the sealed capillaries were placed into the LightCycler rotor.

The inclusion of negative as well as positive controls in each set of experiments is considered to be obligatory in the field of diagnostic PCR. The negative control sample was prepared by replacing the DNA template with PCR grade water. The positive control sample was prepared by adding 2 µl of genomic DNA (approx. 1 ng) of toxigenic *C. diphtheriae* organisms to the master mix.

The following PCR protocol was used for amplification and hybridization probe-based detection of the of the *C. diphtheriae* toxin gene-specific amplicons:

- Denaturation for 10 min at 95°C (to activate the FastStart *Taq* DNA polymerase).
- Amplification

Parameter	Value		
Cycles	50		
Type	None		
	Segment 1	Segment 2	Segment 3
Target temperature [°C]	95	50	72
Incubation time (sec)	10	10	20
Temperature transition rate (°C/sec)	20	20	20
Acquisition mode	None	Single	None
Gains	F1=1; F2=15; F3=30		

- Cooling for 2 min at 40°C

As with every LightCycler experiment, readout of [Red 640] values was performed in channel "F2".

Results

Compared to a well-established *in house* PCR protocol [11] for the detection of toxigenic *C. diphtheriae* organisms in nasopharyngeal swabs (amplification in traditional thermocycler devices and product detection in ethidium bromide-stained agarose gels), the LightCycler system was examined to determine if it could simplify the diagnostic laboratory workflow by reducing the assay's turn-around time, automation, and by reducing the possibility of product contamination frequently associated with post-PCR amplicon manipulation. As with every modified or new PCR protocol in diagnostic microbiology, the sensitivity and specificity of any assay has to be carefully examined before introducing it into routine testing.

Detection Limit of the PCR Assay: Analytical Sensitivity

To determine the assay's lower limit of detection, PCR experiments were performed on serial dilutions of genomic DNA prepared from cultured toxigenic *C. diphtheriae* organisms. Looking at the toxigenic *C. diphtheriae* NCTC 10648 strain, a detection limit of 100 fg of genomic DNA was observed (Figure 2). This corresponds to about 50 genome equivalents. Testing nasopharyngeal swabs spiked with defined numbers of toxigenic *C. diphtheriae* organisms, a detection limit of about 200 organisms per PCR reaction was observed with the toxigenic *C. diphtheriae* NCTC 10648 strain (Figure 3), and a detection limit of about 2000 organisms per PCR reaction was observed with the weak toxigenic *C. diphtheriae* NCTC 3984 strain (Figure 4). Swab samples spiked with defined numbers of the nontoxigenic strain *C. diphtheriae* NCTC 10356 were all found PCR-negative (Figure 5).

The diagnostic sensitivity of the assay was further examined with a dilution series performed on a nasopharyngeal swab specimen obtained from a *C. diphtheriae*-positive patient. After 50 cycles of amplification, positive PCR results were observed even with template DNA prepared from a 1:1000 dilution of the specimen in PBS (data not shown).

Analytical Specificity

The PCR assay was evaluated with 22 clinical samples originating from different patients presenting with diphtheria-like symptoms. Previous PCR testing of the specimens according to the protocol of Nakao and coworkers [11], and conventional culture followed by Elek plate testing, has revealed 3 of 22 samples positive for the *C. diphtheriae* toxin gene. In addition, 3 type strains (toxigenic *C. diphtheriae* NCTC 10648, weak toxigenic *C. diphtheriae* NCTC 3984, and nontoxigenic *C. diphtheriae* NCTC 10356), 18 patient strains of toxigenic *C. diphtheriae*, 2 patient strains of nontoxigenic *C. diphtheriae* as well as 80 isolates of Gram-negative and Gram-positive bacterial species other than *C. diphtheriae* were examined. LightCycler PCR results with the cultured strains and clinical specimens were in perfect agreement with the results of previous PCR testing and all DNA preparations from toxigenic *C. diphtheriae* strains showed amplification by fluorescence, whereas none of the DNA preparations from nontoxigenic *C. diphtheriae* or the other cul-

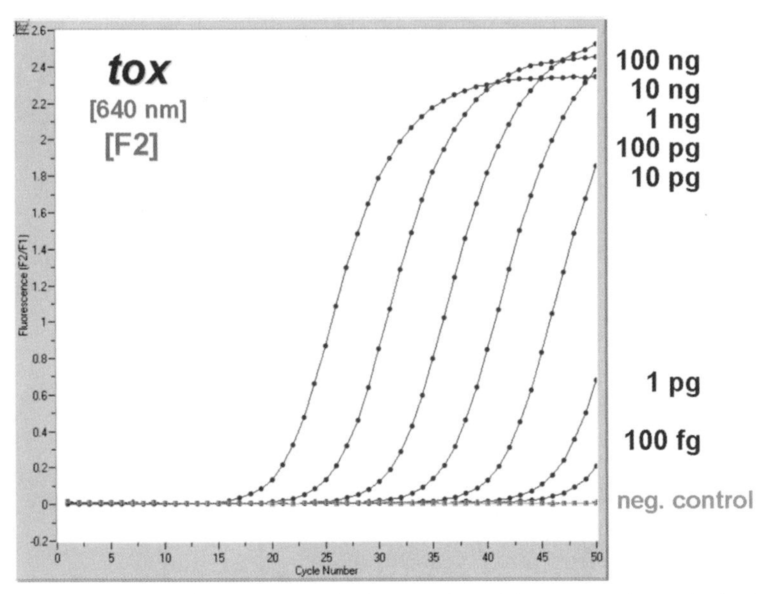

Fig. 2. Sensitivity of the PCR assay for the detection of the *tox* gene determined with serial dilutions of toxigenic *C. diphtheriae* genomic DNA. For each dilution, the corresponding amount of *C. diphtheriae* genomic DNA present in the reaction mixture is given next to the amplicon curve

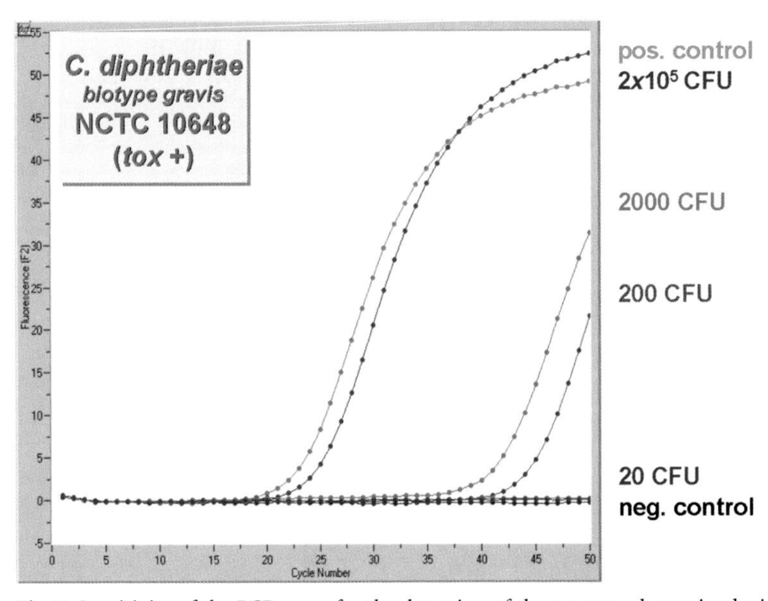

Fig. 3. Sensitivity of the PCR assay for the detection of the *tox* gene determined with nasopharyngeal swabs spiked with defined numbers of toxigenic *C. diphtheriae* NCTC 10648 organisms. For each dilution, the corresponding number of *C. diphtheriae* organisms per PC assay (org./LC) is given next to the amplicon curve

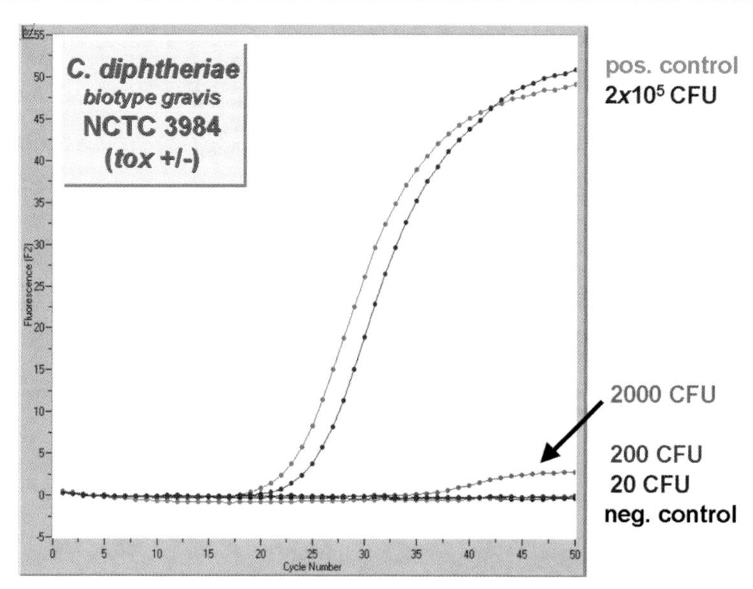

Fig. 4. Sensitivity of the PCR assay for the detection of the *tox* gene determined with nasopharyngeal swabs spiked with defined numbers of weak toxigenic *C. diphtheriae* NCTC 3984 organisms. For each dilution, the corresponding number of *C. diphtheriae* organisms per PCR assay (org./LC) is given next to the amplicon curve

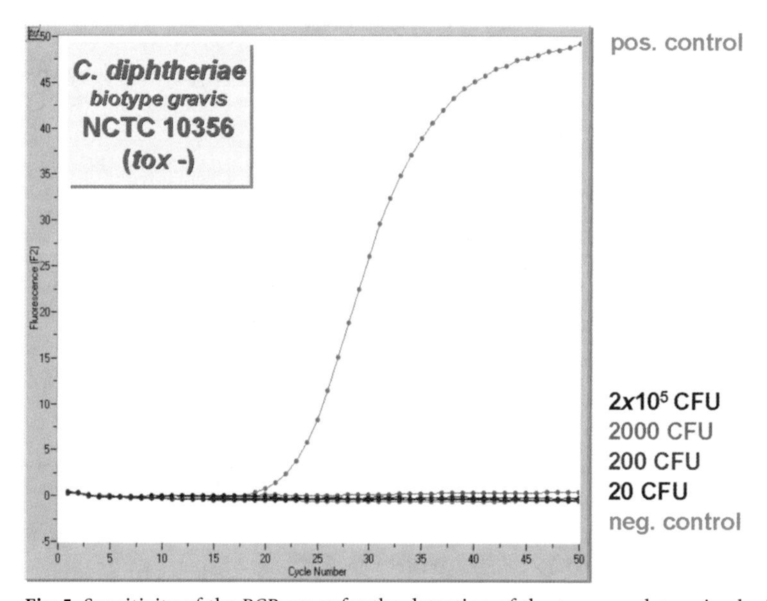

Fig. 5. Sensitivity of the PCR assay for the detection of the *tox* gene determined with nasopharyngeal swabs spiked with defined numbers of nontoxigenic *C. diphtheriae* NCTC 10356 organisms. For each dilution, the corresponding number of *C. diphtheriae* organisms per PCR assay (org./LC) is given next to the amplicon curve

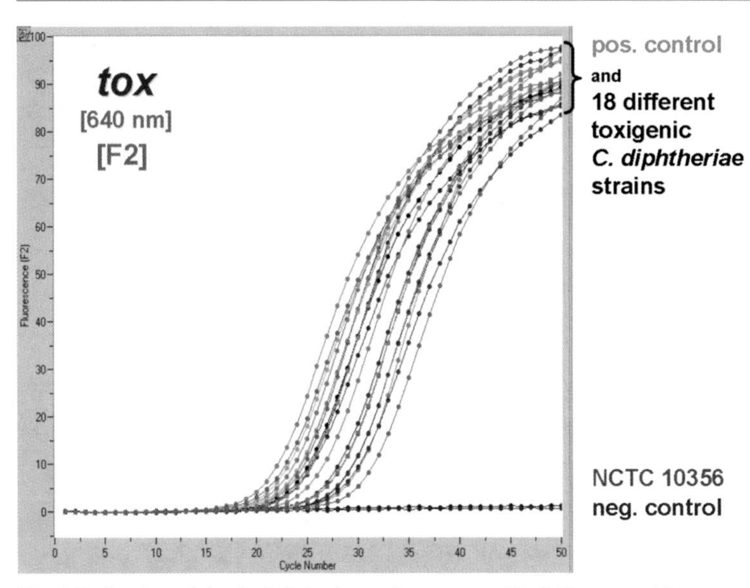

Fig. 6. Evaluation of the *C. diphtheriae* toxin gene-specific PCR assay with a representative set of 19 cultured strains (including nontoxigenic *C. diphtheriae* NCTC 10356 type strain and 18 different patient isolates of toxigenic *C. diphtheriae*)

tured bacterial species examined were recognized. Experimental results on a representative set of toxigenic *C. diphtheriae* patient strains are shown in Fig. 6.

Even after as few as 35 cycles of amplification, the curves generated by the LightCycler software allowed for a clear discrimination between PCR-positive and PCR-negative samples. However, a total of 50 amplification cycles should be performed since also the presence of weak toxigenic *C. diphtheriae* strains could be expected in clinical samples.

Inhibition Events

To investigate PCR inhibition, a separate set of PCR experiments was performed on DNA preparations from 10 nasopharyngeal swab specimens and 10 cultured isolates (spiked with 300 pg of human genomic DNA). Here the primers and probes of the LightCycler Control Kit were used for amlification of the human ß-globin gene. No inhibition events were observed with any of the DNA preparations tested. In contrast to direct testing of clinical specimens known to be prone to inhibition, such as blood or stool samples, there was no apparent inhibition with respiratory specimens or cultured bacteria processed with the HighPure PCR Template Preparation Kit.

Comments

LightCycler PCR

Previous methods for diagnostic PCR have employed a variety of time-consuming and laborious procedures (such as Southern-blot, DNA sequencing, or solid-phase capturing) for sequence-specific characterization of the amplicons generated with clinical specimens. Enhancing the reliability of the results by sequence-specific probes and simplifying the PCR workflow by a completely automated amplification and online detection procedure, the LightCycler system proved itself as a valuable tool for rapid and sensitive identification of toxigenic *C. diphtheriae* organisms in the environment of a routine microbiological laboratory. Once DNA is extracted from suitable specimens or cultured isolates and reaction mixtures are completed, PCR results are available within a period of 60 minutes. Current limitations of the LightCycler system are a total reaction volume of 20 µl (which limits the input of template DNA), and individual sample handling of the reaction cuvettes. An automated platform (MagNA Pure LC instrument) for DNA preparation and completing PCR reaction mixtures has recently become available.

Direct Testing of Clinical Specimens

Due to the rare event of diphtheria infection in Germany at present, only 3 *C. diphtheriae*-positive clinical specimens were available, which were all found positive by LightCycler PCR. A detection limit of at least 50 genome equivalents per PCR reaction, however, should be sensitive enough for the direct identification of toxigenic *C. diphtheriae* organisms in nasopharyngeal or throat swabs. In the course of a prospective study, we will continue to evaluate the diagnostic specificity and sensitivity of our LightCycler assay for direct identification of toxigenic *C. diphtheriae* organisms in clinical specimens.

References

1. Centers for Disease Control and Prevention (1995) Diphtheria epidemic – New independent states of the former Soviet Union, 1990–1994. Morbid Mortal Weekly Rep 44:177–181
2. Centers for Disease Control and Prevention (1995) Diphtheria acquired by U.S. citizens in the Russian Federation and Ukraine-1994. Morbid Mortal Weekly Rep 44:237–244
3. Popovic T, Kim C, Reiss J, Reeves M, Hakao H, Golaz A (1999) Use of molecular subtyping to document long-term persistence of *Corynebacterium diphtheriae* in South Dakota. J Clin Microbiol 37:1092–1099
4. Gubler J, Huber-Schneider C, Gruner E, Altwegg M (1998) An outbreak of nontoxigenic *Corynebacterium diphtheriae* infection: single bacterial clone causing invasive infection among Swiss drug users. Clin Infect Dis 27:1295–1298
5. Funke G, Altwegg M, Frommelt L, von Graevenitz A (1999) Emergence of related nontoxigenic *Corynebacterium diphtheriae* biotype mitis strains in western Europe. Emerg Inf Dis 5:477–480
6. Reinhardt DJ, Lee A, Popovic T (1998) Antitoxin-in-membrane and antitoxin-in-well assays for detection of toxigenic *Corynebacterium diphtheriae*. J Clin Microbiol 36:207–210
7. Pennie RA, Malik AS, Wilcox L (1996) Misidentification of toxigenic *Corynebacterium diphtheriae* as a *Corynebacterium* species with low virulence in a child with endocarditis. J Clin Microbiol 34:1275–1276
8. Pallen MJ, Hay AJ, Puckey LH, Efstratiou A (1994) Polymerase chain reaction for screening clinical isolates of corynebacteria for the production of diphtheria toxin. J Clin Pathol 47:353–356
9. Pallen MJ (1991) Rapid screening for toxigenic *Corynebacterium diphtheriae* by the polymerase chain reaction. J Clin Pathol 44:1025–1026
10. Komiya T, Shibata N, Ito M, Takahashi M, Arakawa Y (2000) Retrospective diagnosis of diphtheria by detection of the *Corynebacterium diphtheriae tox* gene in a formaldehyde-fixed throat swab using PCR and sequence analysis. J Clin Microbiol 38:2400–2402

11. Nakao H, Popovic T (1997) Development of a direct PCR assay for detection of the diphtheria toxin gene. J Clin Microbiol 36:207–210
12. Sulakvelidze A, Kekelidze M, Gomelauri T, Deng Y, Khetsuriani N, Kobaidze K, De Zoysa A, Efstratiou A, Morris JG, Imnadze P (1999) Diphtheria in the republic of Georgia: use of molecular typing techniques for characterization of *Corynebacterium diphtheriae* strains. J Clin Microbiol 37:3265–3270
13. Mikhailovich VM, Melnikov VG, Mazurova IK, Wachsmuth IK, Wenger JD, Wharton M, Nakao H, Popovic T (1995) Application of PCR for detection of toxigenic *Corynebacterium diphtheriae* strains isolated during the Russian diphtheria epidemic, 1990 through 1994. J Clin Microbiol 33:3061–3063
14. Popovic T, Mazurova IK, Efstratiou A, Vuopio-Varkila J, Reeves MW, De Zoysa A, Glushkevich T, Grimont P (2000) Molecular epidemiology of diphtheria. J Infect Dis 181 (Suppl 1):S168–77
15. Nakao H, Pruckler JM, Mazurova IK, Narvskaia OV, Glushkevich T, Marijevski VF, Kravetz AN, Fields BS, Wachsmuth IK, Popovic T (1996) Heterogeneity of diphtheria toxin gene, *tox*, and its regulatory element, *dtx*R, in *Corynebacterium diphtheriae* strains causing epidemic diphtheria in Russia and Ukraine. J Clin Microbiol 34:1711–1716
16. Kaczorek M, Delpeyroux F, Chenciner N, Streeck RE, Murphy JR., Boquet P-LL, Tiollais P (1983) Nucleotide sequence and expression of the diphtheria tox228 gene in *Escherichia coli*. Science 221: 855–858

Rapid Detection of Resistance Associated Mutations in *Mycobacterium tuberculosis* by LightCycler PCR

MARIA J. TORRES*, ANTONIO CRIADO, JOSE C. PALOMARES, JAVIER AZNAR

Introduction

Rifampin (RMP) and isoniazid (INH) are important components of effective multidrug therapy and prophylaxis for *M. tuberculosis* infections. However, widespread use and failure of patients to complete prescribed treatment have led to the emergence of RMP- and INH-resistant strains. RMP resistance is well characterized, and 95% of all *M. tuberculosis* RMP-resistant clinical isolates harbor specific mutations within an 81-bp region of the *rpoB* gene, which is coding for the β subunit of the RNA polymerase [1]. Thus, investigation of RMP resistance is relatively straightforward. In contrast to RMP, genotype testing for INH resistance is much more complex. Several studies have reported alterations in at least four genes, one encoding for catalase-peroxidase (*katG*), the regulatory region of an NADH-dependent enoyl ACP reductase (*inhA*), and a β-Ketoacyl acyl carrier protein synthase (*kasA*), both involved in mycolic acid biosynthesis, and recently the involvement of an alkyl-hydroperoxidase reductase (*ahpC*) has been suggested [2–4]. Although the importance of the *katG* mutations in INH resistance has been clearly demonstrated, the involvement of the other genes is arguable. The highest proportion of mutations is in the *katG* gene, but only some mutations in this gene have been associated with high-level resistance (e.g., deletions or Ser315Thr), whereas others do not appear to confer any resistance to INH (e.g., Arg463Leu).

The need to minimize the transmission of drug-resistant strains requires rapid identification procedures. Due to the slow growth of *M. tuberculosis* organisms, unacceptable delays in the diagnosis of the drug-resistant strains can occur when using conventional cultured-based susceptibility tests. DNA assays have the potential to provide rapid detection of resistance in mycobacteria isolates due to their high sensitivity and specificity. Because these procedures do not rely on in vitro growth, the time between diagnosis and the onset of effective therapy is shortened.

In this report we present a single-tube method for detecting mutations associated with resistance to RMP and INH. It combines both rapid cycle PCR with real-time monitoring of the process and generation of mutation-specific fluorescent probe melting profiles on the LightCycler instrument. Using this method, one strain could

* Maria J. Torres (✉) (e-mail: folia@cica.es)
Departamento de Microbiología, Unidad de Microbiología Molecular, Facultad de Medicina, Universidad de Sevilla, Apartado 914, 41080 Sevilla, Spain

be genotyped within 30 min with no post-PCR sample manipulation. Two fluorescent-labeled hybridization probes recognizing adjacent sequences in the amplicon were present in the reaction mixture. The shorter detection probe (sensor, 5′-LightCycler-Red-640-labeled) covers the predicted site of mutation, while the longer probe (anchor, 3′-fluorescein-labeled) produces the fluorescent signal. After annealing, the fluorophores are in close proximity and there can be a fluorescence resonance energy transfer (FRET) between them, providing real-time monitoring of the amplification process. When PCR was completed, fluorescence was monitored as the temperature was increased through the melting temperature (T_m) of the probe/product duplexes, and a characteristic melting profile for each genotype was obtained. We used four different pairs of probes: (a) Probe 1 designed to detect the two most frequent mutations in the *rpoB* gene related to RMP resistance: Ser531Leu, a change in codon 531 from coding serine to leucine (TCG→TTG), and His526Asp, a change in codon 526 from coding histidine to aspartic acid (CAC→GAC); (b) Probe 2 designed to detect another two mutations in the *rpoB* gene: Gln513Leu, a change in codon 513 from coding glutamine to leucine (CAA→CTA), and Asn518Ser, a change in codon 518 from coding asparagine to serine (AAC→AGC); (c) Probe 3 designed to detect the most frequent mutation (codon 315) related to INH resistance in the *katG* gene; (d) Probe 4 to detect mutations in the regulatory region of the *inhA* gene (nucleotide substitution C209T), also related to INH resistance.

The assay was evaluated with a collection of susceptible and resistant *M. tuberculosis* clinical isolates. The susceptibility testing was performed by the radiometric dilution method, and all the resistant strains were studied by SSCP and sequencing [5, 6].

Materials

Equipment

LightCycler instrument (Roche Diagnostics, Mannheim, Germany)
LightCycler software vers. 3.0 (Roche Diagnostics)

Reagents

Amplification Primers (TIB MOLBIOL, Berlin, Germany)
Hybridization Probes (TIB MOLBIOL)
Chelex 100 (Amersham Pharmacia Biotech, Uppsala, Sweden)

The following reagents were purchased from Roche Diagnostics:
LightCycler DNA Master Hybridization Probes

Procedure

Preparation of Template DNA

A rapid DNA extraction procedure for direct testing of *M. tuberculosis* from Löwenstein-Jensen solid medium was performed as follows: one 10 μl loop of the organisms was resuspended in 100 μl of sterile water and subsequently 100 μl of a Chelex 100 suspension (10%, in water) was added. After mixing thoroughly, the mixture was incubated at 45°C for 45 min, then the suspension was boiled for

5 min, and the bacterial debris was removed by centrifugation (12,000 *g* for 5 min). The supernatant was directly used for amplification.

Primer Design

Three previously published different primer pairs [7] were used for amplifying the *rpoB* gene, the *katG* gene, and the regulatory region of *inhA*.

Hybridization probes were designed to distinguish between different mutations. Details are shown in Table 1. To avoid dimer formation between TR8 and the hybridization probes, a GT at position 3407 (GenBank Accession #L27989) was introduced in the 3′-terminus of the corresponding probe.

LightCycler PCR

The following master mix was used for amplification and hybridization probe-based detection of RMP and INH resistance-associated mutations in *M. tuberculosis* isolates:

	Volume [µl]	[Final]
LightCycler-DNA Master Hybridization Probes	2	1×
$MgCl_2$ stock solution	2.4	4 mM
Primers (10 µM each)	1+1	0.5 µM
Hybridization probes (4 µM each)	1+1	0.1 µM
H_2O (PCR grade)	9.6	
Total volume	18	

To complete the amplification mixture, 18 µl of the master mix and 2 µl of the corresponding template DNA preparation were added to each capillary. After a pulse centrifugation in a microcentrifuge to fill the cuvettes, they were placed into the LightCycler rotor. In each set of experiments we always included negative as well as positive controls. The negative control sample was prepared by replacing the DNA template with PCR-grade water. The positive control sample was prepared by adding 2 µl of genomic DNA (approx. 10 ng/µl) from RMP and INH susceptible *M. tuberculosis* H37Rv to the master mix.

The following PCR protocol was used for amplification:

- Denaturation for 2 min at 95°C
- Amplification

Parameter	Value		
Cycles	30		
Type	None		
	Segment 1	Segment 2	Segment 3
Target temperature [°C]	95	55–60[a]	72
Incubation time [s]	0	5	10
Temperature transition rate [°C/s]	20	20	20
Acquisition mode	None	Single	None
Gains	F1=1; F2=15; F3=30		

[a] The annealing temperature was 55°C for *rpoB* and *inhA*, and 60°C for *katG*.

Table 1. Oligonucleotides

M. tuberculosis rpoB gene (GenBank Accession #L27989)				
	Position	Length	GC (%)	T_m (°C)
Primers				
TR8: GTGCACGTCGCGGACCTCCA	2492R	20	70.0	68.4
TR9: TCGCCGCGATCAAGGAGT	2335	18	61.1	61.0
Product	2335–2492	158		
Probes				
rpo1 anchor: TTCATGGACCAGAACAACCCGCTGTCGGT-F	2379	29	55.2	73.8
rpo1 sensor: LCRed640-ACCCACAAGCGCCGACTGCTGG	2412	22	68.2	70.4
rpo2 anchor: GCTGAGCCAATTCATGGACCAGAACAACC-F	2369	29	51.7	69.8
rpo2 sensor: LCRed640-CTGTCGGGGTTGACCCACAAG-CGC	2400	24	66.7	73.2
M. tuberculosis katG gene (GenBank Accession #X68081)				
Primers				
TB 86: GAAACAGCGGCGCTGATCGT	2759	20	60.0	64.3
TB 87: GTTGTCCCATTTCGTCGGGG	2967R	20	60.0	62.8
Product	2759–2967	209		
Probes				
TB anchor: CGTATGGCACCGGAACCGGTAAGGACGC-F	2886	28	64.3	75.3
TB sensor: LCRed640-TCACCAGCGGCATCGAGGTCGT	2916	22	64.3	69.4
M. tuberculosis regulatory region of fabG *(inhA)* (GenBank Accession #U66801)				
Primers				
TB 92: CCTCGCTGCCCAGAAAGGGA	56	20	65.0	65.2
TB 93: ATCCCCCGGTTTCCTCCGGT	303R	19	68.4	66.5
Product	56–303	248		
Probes				
inh anchor: CCCCTTCAGTGGCTGTTGGCCAGTC-F	226R	23	69.6	68.1
inh sensor: LCRed640-CCCGACAACCTATCATCTCGCC	202R	22	59.0	63.1

- Melting Curve Analysis

Parameter	Value		
Cycles	1		
Type	Melting		
	Segment 1	Segment 2	Segment 3
Target temperature [°C]	95	50	95
Incubation time [s]	30	10	0
Temperature transition rate [°C/s]	20	20	0.1
Acquisition mode	None	None	Cont.
Gains	F1=1; F2=15; F3=30		

- Cooling for 2 min at 40°C

Results

The temperature at which the probes melted from the PCR product during the melting program was calculated using the LightCycler software.

A summary of the T_m for each of the probes together with the changes in T_m for the products derived from resistant and susceptible *M. tuberculosis* strains is shown in Table 2.

We were able to detect the two most frequent mutations responsible for RMP resistance using one of the RMP pair of probes (rpo1 anchor/rpo1 sensor) (Fig. 1). The T_m for the susceptible strain (WT) used as control was 64.3°C, while the change from wild type to mutant at codon 531 (TCG→TTG) resulted in more than 2°C increase in the probe's T_m. We observed a greater than 6°C drop in the probe's T_m in the strains with a mutation at codon 526 (CAC→GAC). A second pair of probes (rpo2 anchor/rpo2 sensor), using the same primers, allowed detection of mutations at codons 513 and 518 (Fig. 2), the T_m for the susceptible strain (WT) was 70.1°C. The change from wild-type to mutant at codon 513 (CAA→CTA) resulted in a 3°C drop in the probe's T_m. Mutation at the codon 518 (AAC→AGC) resulted in a drop of almost 6°C in the T_m. Based on our mutation data in strains isolated in our study area in the last 6 years [5] (and unpublished data), the sensitivity for the detection of RMP resistance would be greater than 95% if we used these two pairs of probes.

Figure 3 shows the results with the INH probe (TB sensor). The T_m for the susceptible strain was 72.8°C. The change from wild type (AGC) to ACC mutant at codon 315 resulted in more than a 3°C drop of melting temperature of the *katG* sensor probe. A drop of almost 5°C in the T_m was found in the strains harboring a different mutation at codon 315 (AGC→AAC).

Figure 4 shows the results with the second probe for INH resistance (inh anchor/inh sensor). The T_m for the susceptible strain (*M. tuberculosis* H37Rv) was 62.7°C.

Nucleotide substitution at position 209 (CT) resulted in a 5°C increase in the probe's T_m.

Table 2. Characteristic melting temperatures for hybridization probes detecting rifampin resistance (*rpo*B) and isoniazid resistance (*kat*G, *inh*A).

Gene	Mutation	T_m±SD[a]
rpoB (rpo1 probes)	WT	64.2±0.4
	531 (TCG→TTG)	66.9±0.5
	526 (CAG→GAC)	58.1±0.2
rpoB (rpo2 probes)	WT	70.1±0.3
	513 (CAA→CTA)	66.4±0.2
	518 (AAC→AGC)	63.3±0.4
katG (TB probes)	WT	72.8±0.5
	315 (AGC→ACC)	68.7±0.5
	315 (AGC→AAC)	67.9±0.1
inhA (inh probes)	WT	62.7±0.4
	C209T	68.5±0.3

[a] Means were derived from at least three determinations.

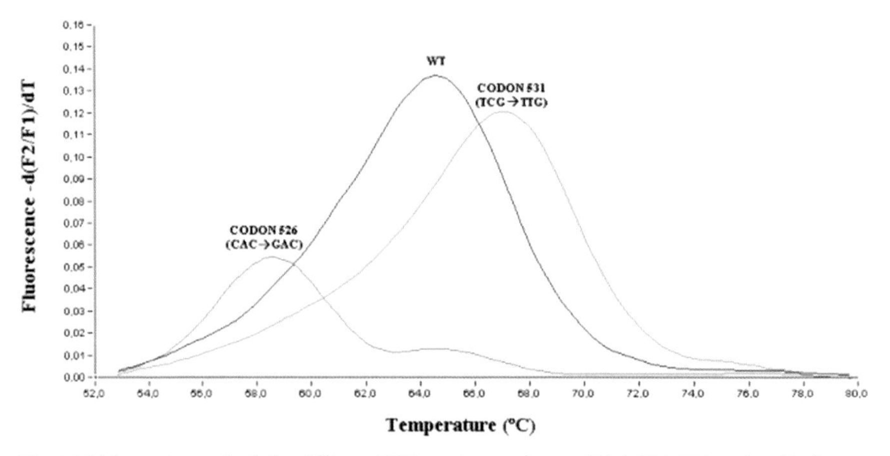

Fig. 1. Melting point analysis for different PCR products using *rpoB* AA 526–531 probes. Each trace shows the changes in fluorescence ratio with time [–(dF2/F1)/dT] with respect to temperature, thus allowing calculation of the temperature at which the probe detaches from the PCR product

The results of this method were validated with selected strains from our *M. tuberculosis* culture collection [5, 6]. We used 20 susceptible and 40 resistant clinical isolates of *M. tuberculosis*. The resistant strains included: 18 mono-RMP-resistant, 18 mono-INH-resistant, and 4 resistant to at least both INH and RMP and then considered as MDR-TB. The sequence inferred from the LightCycler data always agreed with the nucleotide sequencing data [5, 6]. The concordance rate was 100% for each genotype studied.

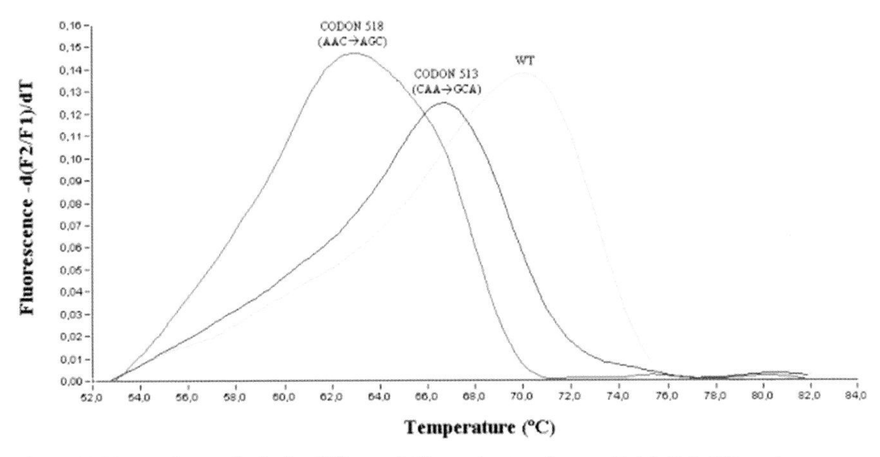

Fig. 2. Melting point analysis for different PCR products using *rpoB* AA 513–518 probes

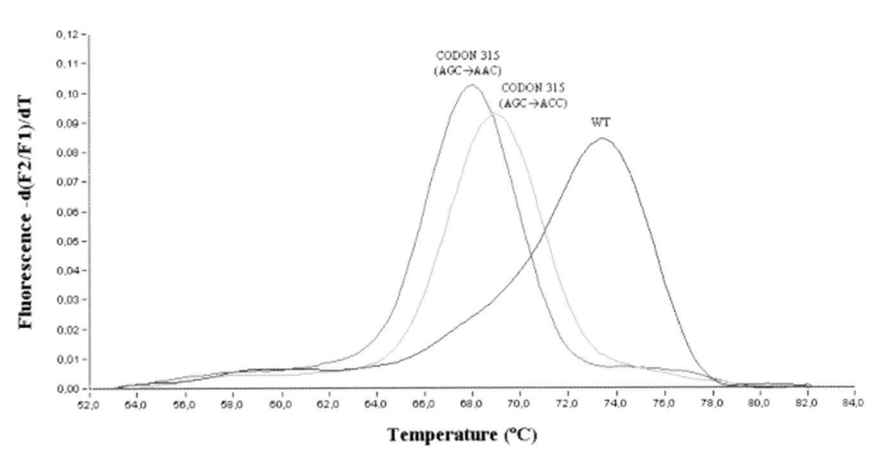

Fig. 3. Melting point analysis for different PCR products using *katG* AA 315 probes

Comments

This new method presents several advantages. First, rapid amplification and analysis allows the test to be completed within 30 min. As the DNA extraction procedure requires only 1 h, the whole detection can be completed in less than 2 h. The procedure can be even simpler using an automated device for template DNA preparation. Second, the LightCycler optical device is capable of measuring fluorescence in two separate channels simultaneously (LCRed640 and LCRed705 fluorophores) and it is possible to use SYBR Green I as a generic donor of FRET (instead of a specific fluorescein-labelled probe), thus allowing analysis of different mutations

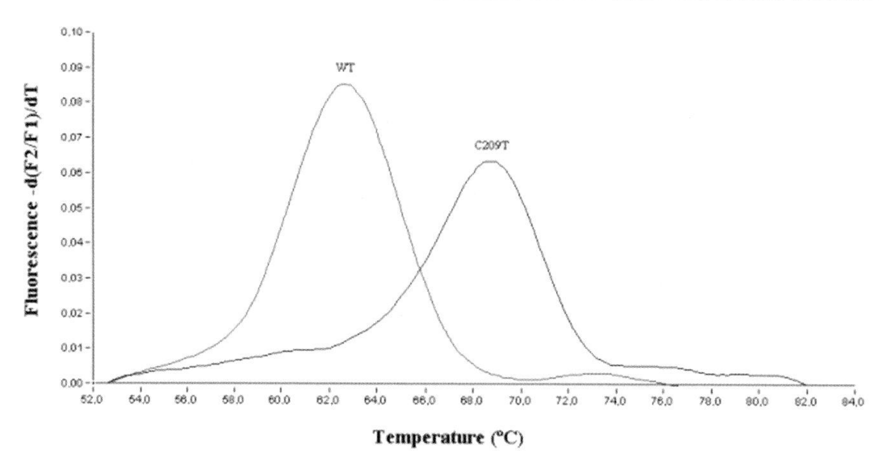

Fig. 4. Melting point analysis for different PCR products using *inhA* Nt 209 probes

within a single test tube. Based on our mutation data in strains isolated in the last 6 years [5] (and unpublished data), when using the additional probe covering codons 513 and 518, the sensitivity for the detection of RMP resistance is greater than 95%, and the reaction may be done in a single tube using the same pair of primers and two different probes, both designed to detect mutations. Third, the assay is run in closed glass capillaries. Postamplification analysis can be performed without opening the capillaries, minimizing the risk of carryover contaminations and overcoming one of the main problems of many amplification tests. Fourth, the LightCycler instrument may be used to determine the genetic make-up of certain bacterial populations composed of subpopulations with different genotypes. Fifth, the method can be adapted to genotyping resistance information emerging from the identification of additional targets, particularly for isoniazid resistance.

One limitation of the LightCycler system is the total volume of 20 µl that can be placed in the capillary tube, which limits the input of template DNA. That would not be a problem working with cultured strains, but could be a limitation when using clinical samples where the number of *M. tuberculosis* bacilli may be very low. Another limitation stems from the theoretical possibility of a natural variability in the sequence where the sensor probes bind, thus leading to a change in the probe T_m with no associated resistance mutation. However, previous studies [8] have suggested that the DNA sequences of *M. tuberculosis* are extraordinarily well preserved and the mutations in the *M. tuberculosis* genome are almost always associated with drug resistance. This fact makes this method even more useful for resistance detection in *M. tuberculosis.*

Finally, we would like to emphasize that this is the first time that real-time PCR has been used to study resistance in *M. tuberculosis* clinical isolates. Reliable results can be obtained faster than with any other current molecular methods and could be applied to clinical samples. In the latter case some modifications may be necessary in the DNA extraction protocol.

References

1. Telenti A, Imboden P, Marchesi F, Lowrie D, Cole S, Coltson MJ, Matter L, Schopfer K, Bodmer T (1993) Detection of rifampin-resistance mutations in *Mycobacterium tuberculosis*. Lancet 341:647–650
2. Banerjee A, Dubnau E, Quemard A, Balasubramanian SK, Um K, Wilson T, Collins D, De Lisle G Jr, Jacobs WR (1994) *Inh*A, a gene encoding a target for isoniazid and ethionamide in *Mycobacterium tuberculosis*. Science 263:227–230
3. Wilson TM, Collins DM (1996) *Ahp*C, a gene involved in isoniazid resistance of the *Mycobacterium tuberculosis* complex. Mol Microbiol 19:1025–1034
4. Zhang Y, Heym B, Allen B, Young D, Cole ST (1992) The catalase-peroxidase gene and isoniazid resistance of *Mycobacterium tuberculosis*. Nature 358:501–593
5. Gonzalez N, Torres MJ, Aznar J, Palomares JC (1999) Molecular analysis of rifampin and isoniazid resistance of *Mycobacterium tuberculosis* clinical isolates in Seville, Spain. Tub Lung Dis 79:187–190
6. Torres MJ, Criado A, Palomares JC, Aznar J (2000) Use of real-time PCR and fluorometry for rapid detection of rifampin and isoniazid resistance-associated mutations in *Mycobacterium tuberculosis*. J Clin Microbiol 38:3194–3199
7. Telenti A, Honoré N, Bernasconi C, March J, Ortega A, Takiff HE, Cole ST (1997) Genotyping assessment of isoniazid and rifampin resistance in *Mycobacterium tuberculosis*: a blind study at reference laboratory level. J Clin Microbiol 35:719–723
8. Sreevatsan S, Pan X, Stockbauer KE, Connell ND, Kreiswirth BN, Whittam TS, Musser JM (1997) Restricted structural gene polymorphism in the *Mycobacterium tuberculosis* complex indicates evolutionary recent global dissemination. Proc Natl Acad Sci USA 74:9869–9874

Duplex LightCycler PCR Assay for the Rapid Detection of Methicillin-Resistant *Staphylococcus aureus* and Simultaneous Species Confirmation

Udo Reischl*, Hans-Jörg Linde, Birgit Leppmeier, and Norbert Lehn

Introduction

Staphylococcus aureus has been well known as a major pathogen for many years, causing nosocomial and community-acquired infections. Clinical presentations include superficial, deep-skin, and soft tissue infections, pneumonia, endocarditis, osteomyelitis, bacteremia with metastatic abscess formation, as well as a variety of toxin-mediated diseases including gastroenteritis, staphylococcal scalded skin syndrome, and toxic shock syndrome. Depending on the results of susceptibility testing, penicillin or β-lactamase-stable β-lactams are considered to be the drugs of choice for treatment in serious *S. aureus* infections [11]. Since the introduction of semi-synthetic penicillins (such as oxacillin) to clinical use, the occurrence of methicillin-resistant strains of *S. aureus* (MRSA) has increased steadily and nosocomial infections caused by such isolates have become a serious problem worldwide with varying degrees of prevalence [1, 16, 24]. Bacteria have surface membrane localized proteins with binding affinity for β-lactam antibiotics called penicillin-binding proteins (PBPs) which are expressed constitutively. Methicillin-resistance in *S. aureus* is associated with the production of an additional penicillin-binding protein PBP2a (encoded by the *mecA* gene), with a low binding affinity for β-lactam antibiotics. Methicillin resistance is complicated by most of these strains being resistant to other classes of antibiotics. In rare cases methicillin resistance may also be associated with mechanisms independent of *mecA*, such as the hyperproduction of β-lactamase or the production of methicillinases, resulting in borderline resistance.

The detection of methicillin resistance has important implications for treatment and management of the patient. At present, glycopeptides are recommended for the therapy of infections caused by MRSA [11]; however, strains with only intermediate susceptibility to vancomycin have been isolated in Japan and the USA [22]. In addition, an extensive set of hygienic precautions must be taken to limit the spread of MRSA [8, 25].

In the clinical laboratory, *S. aureus* is identified by growth characteristics, β-hemolysis on blood agar plates, and the subsequent detection of catalase and

* Udo Reischl (✉) (e-mail: Udo.Reischl@klinik.uni-regensburg.de)
Institute of Medical Microbiology and Hygiene, University of Regensburg, Franz-Josef-Strauß-Allee 11, 93053 Regensburg, Germany

coagulase activity or specific surface constituents. The deoxyribonuclease and thermostable endonuclease tests have been used as confirmatory tests in cases of inconclusive or negative coagulase tests. Conventional susceptibility testing of *S. aureus* detects resistance to methicillin or oxacillin if agar dilution or agar screen methods are performed according to NCCLS standards [10]. However, standard susceptibility tests are time-consuming and since the phenotypic expression of methicillin resistance in vitro is found to be heterogeneous and sometimes difficult to induce, false-negative results may be observed. A number of misclassifications of resistance with automated susceptibility testing systems or of species with commercially available latex agglutination kits have been reported recently [14, 19, 20, 21, 26]. Therefore the rapid and sensitive detection of the *mecA* gene by means of in vitro nucleic acid amplification has evolved as the method of choice for rapid and definitive identification of MRSA and other staphylococci [2–4, 6, 7, 13, 23]. Since *S. aureus* can resemble coagulase-negative staphylococci (CoNS) on visual examination of agar plates and the coagulase status is not always easily established in a timely fashion, combined testing for *mecA* and a *S. aureus*-specific marker is attractive [9].

Here we describe the development of a rapid and reliable duplex PCR assay for the simultaneous detection of a recently described *S. aureus*-specific genomic fragment [12] and the *mecA* gene designed to meet these requirements in routine testing. Since amplification and sequence-specific detection of the amplicons were performed with the LightCycler device for ultrarapid thermal cycling and using the dual-color option, a total assay time of less than 50 min was achieved.

The assay was evaluated with a collection of 165 well-characterized clinical isolates of *S. aureus*, including 55 MRSA isolates with different pulsed-field gel electrophoresis patterns, 185 CoNS, as well as 80 isolates of Gram-negative and Gram-positive bacterial species other than *Staphylococcus*. PCR results obtained with the 430 bacterial strains tested were in perfect agreement with the results of conventional microbiological testing.

Materials

Equipment

LightCycler instrument (Roche Diagnostics, Mannheim, Germany)
LightCycler software vers. 3.3
Oligo Primer analysis software (Molecular Biology Insights, Inc., Cascade, Colo., USA)

Reagents

Amplification Primers (Metabion, Munich, Germany)
Hybridization Probes (TIB MOLBIOL, Berlin, Germany)

The following reagents were purchased from Roche Diagnostics:
High-Pure PCR Template Purification Kit
LightCycler-DNA Master Hybridization Probes
LightCycler Control Kit DNA

Procedure

Bacterial Strains and Susceptibility Testing

The bacterial isolates used in this study comprised 5 type strains of methicillin-susceptible *S. aureus* (MSSA) (ATCC 35696, ATCC 27704, ATCC 23361, ATCC 15752, and ATCC 27695), and 5 MRSA (ATCC 43300, ATCC 33591, ATCC 33592, ATCC 33593, and ATCC 33594), as well as 155 clinical isolates of *S. aureus*, including 55 MRSA isolates with different patterns in pulsed-field gel electrophoresis. Other clinical isolates consisted of 185 CoNS, including methicillin-resistant isolates of *S. epidermidis* (n=3), *S. hominis* (n=2), and *S. pasteuri* (n=1) as well as 179 methicillin-susceptible isolates of *S. epidermidis*, *S. hominis*, *S. warneri*, *S. capitis*, and *S. haemolyticus*, and members of 40 Gram-negative and 40 Gram-positive bacterial species.

The strains were maintained on Columbia blood agar. *S. aureus* was identified by colony morphology, presence of gram-positive cocci in clumps, and positive catalase reaction. Coagulase production was demonstrated using a commercial latex agglutination test system (Slidex Staph kit, bioMerieux, Marcy l'Etoile, France). Oxacillin susceptibility was determined by the agar screen method with Mueller-Hinton agar containing 2% NaCl and 6 mg/l of oxacillin for *S. aureus*, or 0.5 mg/l of oxacillin, respectively, for CoNS [10].

For analytical sensitivity determinations, a subculture of MRSA strain ATCC 33592 in the logarithmic phase of growth (optical density at 600 nm $\approx$ 0.7) was diluted serially in phosphate-buffered saline. As described previously [12], each dilution was tested in PCR assays to determine the minimal number of CFUs which could be detected. Detection limits in terms of numbers of CFUs were estimated by a standard plating procedure.

Preparation of Template DNA

Bacterial DNA was prepared, using the High Pure PCR Template Preparation Kit according to the manufacturer's instructions. Briefly, clinical specimens were cultured on blood agar plates overnight at 37°C and portions of individual bacterial colonies were suspended in 200 µl of PBS buffer. In total, 15 µl of a lysozyme solution (10 mg/ml in Tris-HCl, pH 8.0) was added to the suspension and incubated at 37°C for 10 min. After adding 200 µl of binding buffer, 40 µl of a proteinase K solution (20 mg/ml) and a further incubation at 70°C for 10 min, 100 µl of isopropanol was added and the mixture was applied to the High Pure spin column. Following the centrifugation and wash steps, bacterial DNA was eluted with 200 µl of elution buffer and a 2-µl aliquot was directly transferred to PCR. The remainder was stored at –20°C for further experiments.

Alternatively, portions of individual bacterial colonies were suspended in 100 µl of lysis buffer containing 1% Triton X-100, 0.5% Tween 20, 10 mM Tris-HCl, pH 8.0, and 1 mM EDTA [17], and incubated in a screw-capped reaction tube for 10 min in a boiling water bath. After a 2-min centrifugation step at 10,000 *g* to sediment the debris, a 2-µl aliquot of the clear supernatant was directly transferred to PCR. Stored at –20°C, the remainder should be used for further experiments within several days.

Primer and Hybridization Probe Design

Nucleotide sequences of the oligonucleotide primers and fluorescence-labeled hybridization probes, designed for amplification and sequence-specific detection of a 180-bp fragment within the *mecA* gene, and a 180-bp fragment within a *S. aureus*-specific genomic marker [12], respectively, are listed in Table 1.

The selection of candidate sequences for LightCycler hybridization probes was aided by the Oligo software in order to obtain comparable T_m, GC contents within the range of 35%–60%, and to avoid stable secondary structures, regions of significant self-complementarity, stretches of palindromic sequences, and close proximity between the primers and the probes. Annealing sites of primers and hybridization probes within the corresponding GenBank sequences are depicted in Figs. 1 and 2.

Table 1. Oligonucleotides

S. aureus mecA gene for penicillin-binding protein 2a (GenBank Accession #X52593)			
	Length	GC (%)	T_m (°C)
Primers			
GGTGAAGATATACCAAGTGATTA	23	34.8	58.4
GTGAGGTGCGTTAATATTGC	20	45.0	60.1
Probes			
CAGGTTACGGACAAGGTGAAATACTGATT-F	29	41.4	65.7
LCRed640-ACCCAGTACAGATCCTTTCAATCTATAGCG	30	43.3	66.4
S. aureus strain ATCC 25923 clone pSa-442 *Sau*3AI fragment (GenBank Accession #AF033191)			
Primers			
GTCGGTACACGATATTCTTCACG	23	47.8	62.3
CTCTCGTATGACCAGCTTCGGTAC	24	54.2	65.4
Probes			
TACTGAAATCTCATTACGTTGCATCGGAA-F	29	37.9	65.6
LCRed705-ATTGTGTTCTGTATGTAAAAGCCGTCTTG	29	37.9	65.0

```
                      Primer >
PCR Prod     1   ggtgaagatataccaagtgattatccattttataatgctcaaatttcaaacaaaaattta 60
X52593    1605   ............................................................ 1664

                                                               F LCRed640
PCR Prod    61   gataatgaaatattattagctgatt caggttacggacaaggtgaaatactgatt aaccca 120
X52593    1665   ............................................................ 1724

                                                      < Primer (antisense)
                                                      cgttataattgcgtggagtg
PCR Prod   121   gtacagatcctttcaatctatagcgcattagaaaataatggcaatattaacgcacctcac 180
X52593    1725   ............................................................ 1784
```

Fig. 1. Alignment of the amplified fragment within the *S. aureus mecA* gene for penicillin-binding protein 2a (GenBank Accession #X52593). Primer annealing sites, orientation, and the sequence of *mecA*-specific hybridization probes are indicated by **bold** and underlined letters

```
                    Primer >
PCR Prod    1       gtcggtacacgatattcttcacgactaaataaacgctcattcgcgattttataaatgaat 60
AF033191   12       ............................................................ 71

                                                                        F  LCRed705
PCR Prod   61       gttgataacaatgttgtattatc tactgaaatctcattacgttgcatcggaa acattgtg 120
AF033191   72       ............................................................ 131

                                                         < Primer (antisense)
                                                       catggcttcgaccagtatgctctc
PCR Prod  121       ttctgtatgtaaaagccgtcttgataatctttagtagtaccgaagctggtcatacgagag 180
AF033191  132       ............................................................ 191
```

Fig. 2. Alignment of the amplified fragment within the *S. aureus* clone pSa-442 *Sau*3AI fragment (GenBank Accession #AF033191). Primer annealing sites, orientation, and the sequence of pSa442-specific hybridization probes are indicated by *bold* and *underlined letters*

Inhibition Control

The LightCycler Control Kit DNA was used for the detection of *Taq* DNA polymerase inhibitors, possibly present in DNA preparations from clinical samples. The latter kit provides control DNA template, primers, and hybridization probes for amplifying and specific detection of the human β-globin gene. To identify even weak inhibition events, 300 pg of human genomic DNA was amplified in each case.

LightCycler PCR

The following master mix was used for amplification and hybridization probe-based detection of the *mecA*- and *S. aureus*-specific amplicons:

	Volume [μl]	[Final]
LightCycler-DNA Master Hybridization Probes	2	1×
$MgCl_2$ stock solution (25 mM)	3.2	5 mM
Primers *mecA* (10 μM each)	1+1	0.5 μM each
Primers *Sa442* (10 μM each)	0.5+0.5	0.25 μM each
Hybridization probes (2 μM each)	2+2+2+2	0.2 μM each
H_2O (PCR grade)	1.8	
Total volume	18	

To complete the amplification mixtures, 18 μl of master mix and 2 μl of the corresponding template DNA preparation were added to each capillary. After a short centrifugation, the sealed capillaries were placed into the LightCycler rotor.

The inclusion of negative as well as positive controls in each set of experiments is considered to be obligatory in the field of diagnostic PCR. The negative control sample was prepared by replacing the DNA template with PCR grade water. The positive control sample was prepared by adding 2 μl of MRSA genomic DNA (approximately 1 ng) to the master mix.

The following PCR protocol was used for amplification and hybridization probe-based detection of the *mecA*- and *S. aureus*-specific amplicons:

- Denaturation for 2 min at 95°C
- Amplification

Parameter	Value		
Cycles	50 (or 40)		
Type	None		
	Segment 1	Segment 2	Segment 3
Target temperature [°C]	95	50	72
Incubation time [s]	0	10	20
Temperature transition rate [°C/s]	20	20	20
Acquisition mode	None	Single	None
Gains	F1=1; F2=15; F3=30		

- Cooling for 2 min at 40°C

As with every LightCycler dual color experiment, readout of LCRed640 values was performed in channel F2, and readout of LCRed705 values was performed in channel F3.

Results

Based on established PCR protocols for the detection of MRSA organisms, the LightCycler system was examined to determine if it could simplify the diagnostic laboratory workflow by multiplexing, automation, and reducing the assay's turn-around time.

Channel Crosstalk

The "dual color" option of the LightCycler software allows simultaneous measurement of fluorescence values at wavelengths of 640 nm and 705 nm, respectively. This allows the use of multiplex PCR assays with the analysis of two parameters in a single tube. As with every fluorescence-based multiparameter assay, the potential crosstalk between different channels of the wavelength spectrum is one of the most critical parameters. To determine the specificity of this particular LightCycler feature, we analyzed the generation of *mecA*-specific and *S. aureus*-specific amplicons with genomic DNA of an MRSA strain. When fluorescence values were quantitatively measured at the respective wavelengths and displayed separately; no significant crosstalk between the channels was observed (data not shown).

Detection Limit of the PCR Assay: Analytical Sensitivity

PCR experiments performed on MRSA strain ATCC 33592, grown and serially diluted in phosphate-buffered saline as described previously [12], revealed a detection limit of approximately 50 CFU (50 genome equivalents) per assay for the *mecA*-specific amplification (see Fig. 3a). For amplification and detection of the *S. aureus*-specific genomic fragment, an identical detection limit of approximately 50 CFU of *S. aureus* per assay was determined (see Fig. 3b). The duplex approach, containing four different primer oligonucleotides and four different hybridization probes within a single capillary, revealed identical detection limits compared to testing in separate capillaries (data not shown). Significant formation

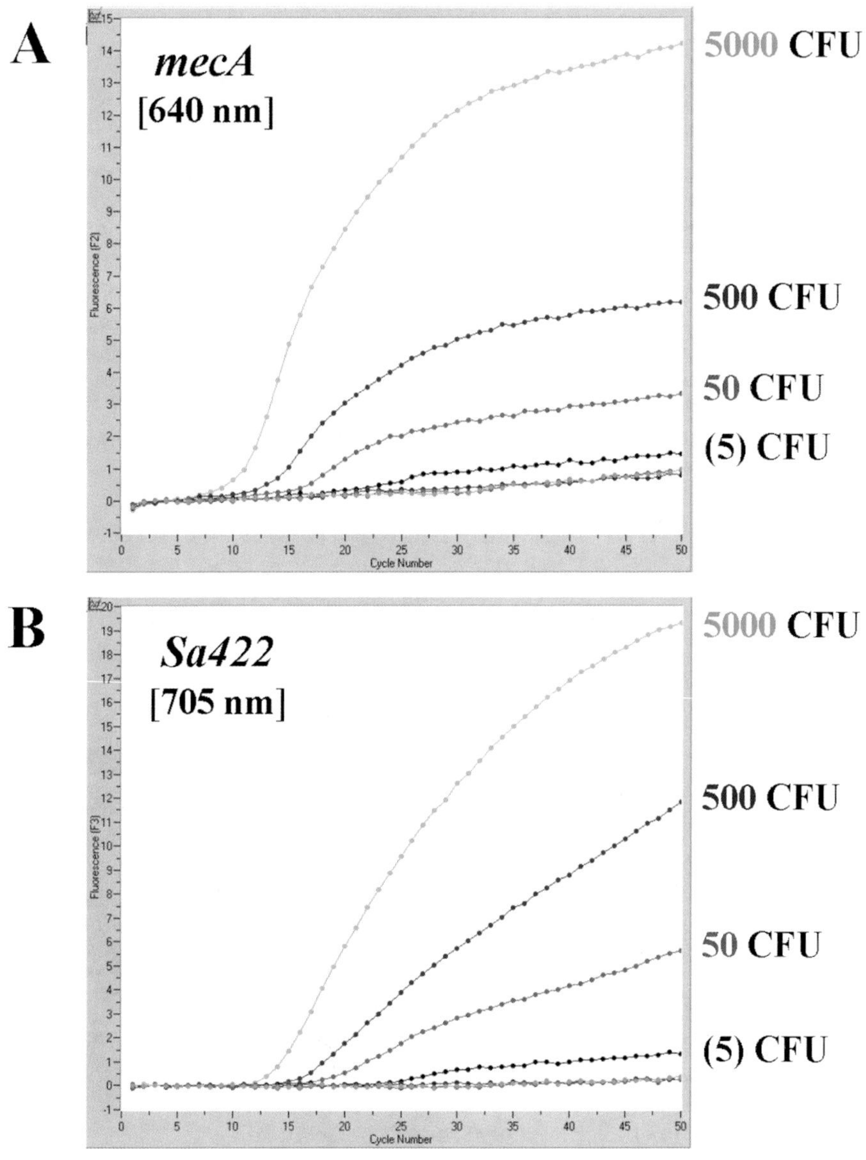

Fig. 3. Sensitivity of the PCR assay for the detection of the *mecA* gene (**a**) and the *S. aureus*-specific chromosomal DNA fragment (**b**), determined with serial dilutions of the MRSA ATCC 33592 strain. For each dilution, the corresponding number of CFUs is given next to the amplicon curve. The numbers of CFUs in *parentheses* indicate a subsequent dilution of the respective DNA preparation

of primer dimers, secondary structures, or other cross-reactions between oligonucleotide components, which frequently interfere with the analytical sensitivity of multiplex PCR approaches, were not apparent in this particular assay.

Specificity

A total of 10 reference strains of *S. aureus* and 155 clinical isolates of *S. aureus* were used as templates for the duplex PCR assay. The results on a representative panel of strains are depicted in Fig. 4. Visual examinations of the plots generated by the LightCycler software (cycle number vs fluorescence values of individual capillaries) allowed for a clear discrimination between positive and negative samples.

All 165 strains were amplified and detected with the *S. aureus*-specific set of primers and hybridization probes, and all 60 methicillin-resistant strains were correctly identified with the *mecA*-specific set of primers and hybridization probes. Compared with the results of conventional identification and susceptibility testing, the PCR assay showed 100% specificity and sensitivity for species identification and detection of methicillin-resistance for the 165 *S. aureus* isolates investigated.

Specificity was further confirmed by testing a panel of 185 CoNS, including methicillin-resistant strains, as well as members of 40 Gram-negative and 40 Gram-positive bacterial species other than *Staphylococcus*. Among the CoNS, methicillin-resistant isolates were positive for the *mecA* gene but negative for the *S. aureus* marker (e.g., the negative curve in Fig. 3b represents a methicillin-resistant *S. epidermidis*). All other bacterial isolates were found negative for both parameters of the duplex PCR assay. These findings for the *S. aureus* marker are in accordance with the results of Martineau et al. [12], who reported 100% sensitivity and specificity for this particular *S. aureus* target sequence.

Inhibition Events

To investigate the possibility of *Taq* DNA polymerase inhibition, 64 samples were randomly selected from *S. aureus* DNA preparations obtained with both extraction methods (High Pure PCR Template Preparation Kit and boiling procedure, respectively). Each of the samples was spiked with 300 pg of human genomic DNA and PCR was performed using primers and hybridization probes of the LightCycler Control Kit, allowing for a specific detection of the human β-globin gene (with 30 pg of human genomic DNA as the lower limit of detection). No inhibition was observed in any of the DNA preparations tested. In contrast to direct testing of clinical specimens, where the presence of compounds inhibitory for the *Taq* DNA polymerase is frequently observed, this issue seemed to be less problematic with bacterial colonies taken from blood agar plates.

Comments

LightCycler PCR

Previous methods for diagnostic PCR have used a variety of time-consuming and laborious procedures (such as Southern blot, DNA sequencing or solid-phase capturing) for sequence-specific characterization of the amplicons generated from clinical specimens. Enhancing the reliability of the results by

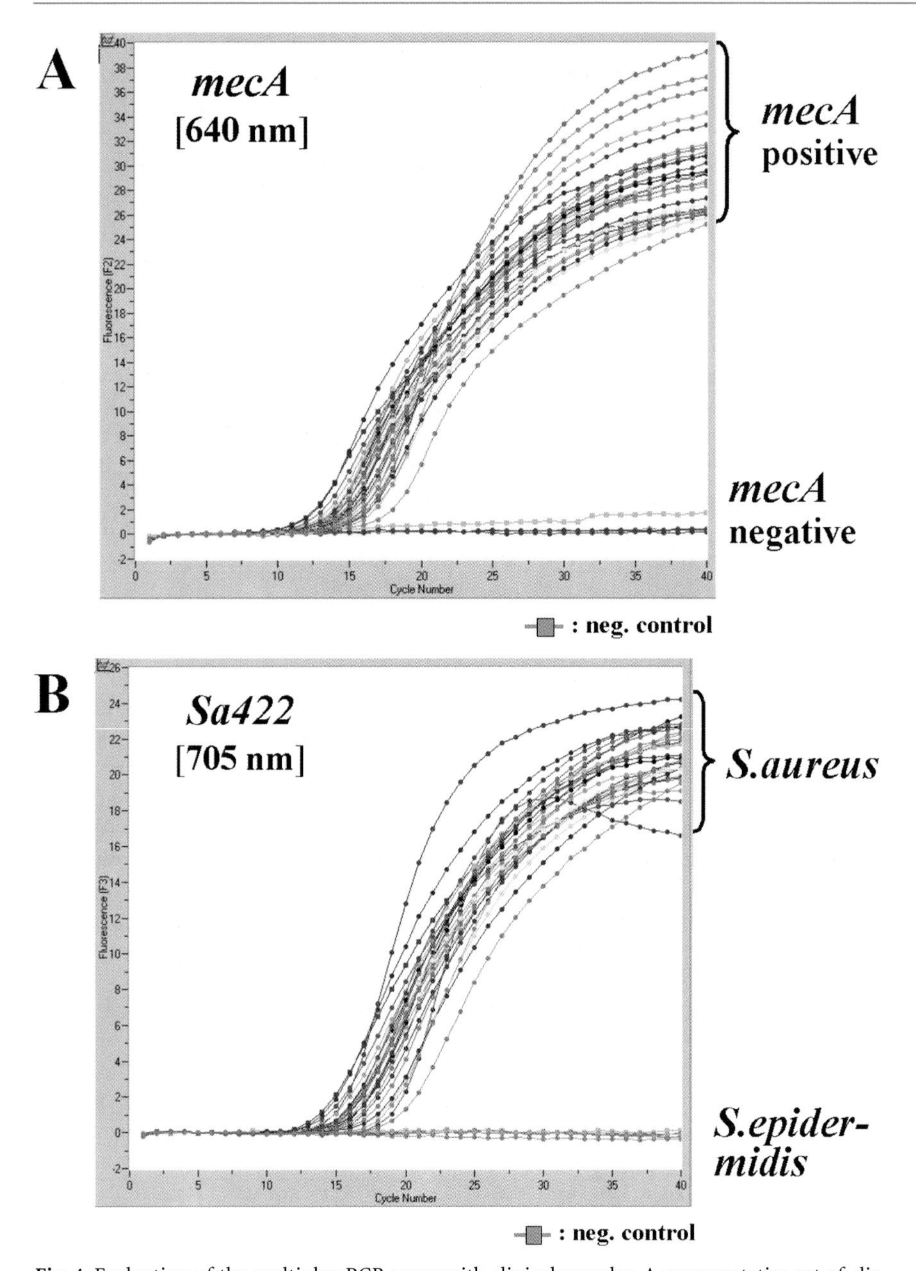

Fig. 4. Evaluation of the multiplex PCR assay with clinical samples. A representative set of clinical samples was simultaneously tested for the *mecA* gene (**a**) and a *S. aureus*-specific genomic fragment *Sa442* (**b**). The depicted curves representing *mecA*-positive isolates are indicated by *parentheses*. The *mecA*-negative samples within this panel were two methicillin-sensitive strains of *S. aureus*. The *Sa442*-negative samples in this panel were a methicillin-resistant strains of *S. epidermidis*

sequence-specific hybridization probes and simplifying the PCR workflow by a completely automated amplification and online detection procedure, the LightCycler system proved itself as a valuable tool for rapid identification of MRSA in the environment of a routine microbiological laboratory setting. Once growth of staphylococci is observed on agar plates, a portion of the colony can be transferred to PCR and, after 10 min of physical manipulation necessary for DNA extraction and completion of the reaction mixture, specific PCR results are available within 1 h.

DNA Extraction from Cultured Bacteria

A number of sophisticated protocols have been published for the extraction of genomic DNA from staphylococci [15]. Most of them are intended for the preparation of intact chromosomal DNA for subsequent genotyping procedures such as arbitrarily primed PCR or pulsed-field gel electrophoresis, which usually have high demands on the quality of input DNA. As the PCR is usually less fastidious about DNA quality, two rapid protocols for the preparation of template DNA from cultured staphylococci were compared.

Starting with a cultured MRSA strain, PCR experiments were performed on serial dilutions of DNA preparations obtained with the High Pure PCR Template Preparation Kit and the boiling procedure. Almost identical quantitative PCR results were observed with both methods for the extraction of amplifiable template DNA from cultured staphylococci.

Duplex PCR Assays

It is well known that individual amplification efficiencies may vary in a multiplex PCR reaction. The production of primer-primer or primer-probe artifacts can be increased when combining more than one set. This effect can be minimized by careful selection of the oligonucleotide sequences. The effective concentrations of oligonucleotide components and the different amplification efficiencies of individual target sequences also influence the overall performance of a multiplex PCR approach. Although we invested a lot of effort in the design of suitable primer and hybridization probe sequences, some deleterious effects with respect to the sensitivity of *mecA* detection were experienced in the evaluation phase with certain batches of primer oligonucleotides ordered from different suppliers.

Systematic investigation of the assay conditions revealed that amplification of the *Sa442* target sequence was much more efficient than the amplification of the *mecA* target sequence (the *mecA* gene is known to be a difficult target for PCR amplification). A strategy to obtain balanced conditions in multiplex assays is the dilution of the superior primer pair [5]. Therefore, should any *mecA* sensitivity problems be observed, the sensitivity of the duplex PCR assay could be easily restored by reducing the total concentration of *Sa442* primers in the PCR reaction mixture (e.g., to a final concentration of 0.1 μM). Once a suitable concentration of *Sa442* primers was determined, the duplex LightCycler PCR assay was found to be highly reproducible in consecutive experiments with large series of clinical specimens. Compared to a previously published protocol [18], we have also reduced the size of the *mecA* amplicon from 409 bp down to 180 bp in order to raise the amplification efficiency of this particular target. This should further minimize the probability of false-negative *mecA* PCR results.

Fast Start DNA Master Hybridization Probes

Generally, it has been our experience that the analytical sensitivity of LightCycler PCR assays can be raised with the use of LightCycler Fast Start DNA Master Hybridization Probes. This observation was confirmed in the case of the presented MRSA dual color assay (detection limit of approximately 5 CFUs of MRSA per assay; data not shown). In the course of comparative experiments with LightCycler-DNA Master Hybridization Probes and LightCycler Fast Start DNA Master Hybridization Probes, however, an overall reduction in fluorescence signal intensity was observed using the Fast Start kit. Taking into account that the PCR assay should be used with bacterial colonies (where sensitivity is not a major issue), we recommend the use of the LightCycler-DNA Master Hybridization Probes for routine applications of the presented protocol.

Direct Testing of Clinical Specimens

A detection limit of at least 50 genome equivalents of MRSA per PCR reaction is considered to be more than sufficient for a culture-confirmation assay. In combination with an appropriate DNA extraction protocol, this assay should also be sensitive enough for the direct identification of *mecA*-positive *S. aureus* organisms in clinical specimens such as swabs or blood culture samples. From the diagnostic point of view, however, direct detection of MRSA should be restricted to clinical specimens from normally sterile sites, where merely the presence of a single bacterial species could be expected. If a clinical specimen contains, for example, a mixture of MSSA and methicillin-resistant CoNS (even in trace amounts in cases of contamination), any PCR assay performed directly on this particular specimen will be positive for both *S. aureus* (due to the presence of MSSA) and *mecA* target sequences (due to the presence of the *mecA* gene in the methicillin-resistant CoNS). Under this combination of circumstances, which are not expected to be rare in the case of testing clinical specimens such as throat or nasal swabs, the positive PCR results are simulating the presence of MRSA organisms. A serious consequence of these false-positive results could be the recommendation of vancomycin for specific therapy (which is definitely unnecessary in this particular situation). Due to this dilemma, we strongly recommend the application of MRSA PCR assays with bacterial colonies on agar plates judged to be pure. A negative *mecA* PCR result, however, proves the absence of MRSA organisms in the examined clinical specimen and could be helpful in assessing the treatment and management of a given patient.

References

1. Barber M (1961) Methicillin-resistant staphylococci. J Clin Pathol 14:385–393
2. Bekkaoui F, McNevin JP, Leung CH, Peterson GJ, Patel A, Bhatt RS, Bryan RN (1999) Rapid detection of the *mecA* gene in methicillin-resistant staphylococci using a colorimetric cycling probe technology. Diagn Microbiol Infect Dis 34:83–90
3. Chambers HF (1997) Methicillin resistance in staphylococci: molecular and biochemical basis and clinical implications. Clin Microbiol Rev 10:781–791
4. Cloney L, Marlowe C, Wong A, Chow R, Bryan R (1999) Rapid detection of *mecA* in methicillin resistant *Staphylococcus aureus* using cycling probe technology. Mol Cell Probes 13:191–197

5. Henegariu O, Heerema NA, Dlouhy SR, Vance GH, Vogt PH (1997) Multiplex PCR: critical parameters and step-by-step protocol. BioTechniques 23:504–511
6. Hussain Z, Stoakes L, Lannigan R, Longo S, Nancekivell B (1998) Evaluation of screening and commercial methods for detection of methicillin resistance in coagulase-negative staphylococci. J Clin Microbiol 36:273–274
7. Hussain Z, Stoakes L, Massey V, Diagre D, Fitzgerald V, El Sayed S, Lannigan R (2000) Correlation of oxacillin MIC with *mecA* gene carriage in coagulase-negative staphylococci. J Clin Microbiol 38:752–754
8. Jernigan JA, Titus MG, Groschel DA, Getchell-White S, Farr BM (1996) Effectiveness of contact isolation during a hospital outbreak of methicillin-resistant *Staphylococcus aureus*. Am J Epidemiol 143:496–504
9. Kearns AM, Seiders PR, Wheeler J, Freeman R, Steward M (1999) Rapid detection of methicillin-resistant staphylococci by multiplex PCR. J Hosp Infect 43:33–37
10. Kohner P, Uhl J, Kolbert C, Persing D, Cockerill F (1999) Comparison of susceptibility testing methods with *mecA* gene analysis for determining oxacillin (methicillin) resistance in clinical isolates of *Staphylococcus aureus* and coagulase-negative *Staphylococcus* spp. J Clin Microbiol 37:2952–2961
11. Lowy FD (1998) *Staphylococcus aureus* infections. N Engl J Med 339:520–532
12. Martineau F, Picard FJ, Roy PH, Ouellette M, Bergeron MG (1998) Species-specific and ubiquitous-DNA-based assays for rapid identification of *Staphylococcus aureus*. J Clin Microbiol 36:618–623
13. Murakami K, Minamide W, Wada K, Nakamura E, Teraoka H, Watanabe S. (1991) Identification of methicillin-resistant strains of staphylococci by polymerase chain reaction. J Clin Microbiol 29:2240–2244
14. Nakatomi Y, Sugiyama J (1998) A rapid latex agglutination assay for the detection of penicillin-binding protein 2'. Microbiol Immunol 42:739–743
15. Olmos A, Camarena JJ, Nogueira JM, Navarro JC, Risen J, Sanchez R (1998) Application of an optimized and highly discriminatory method based on arbitrarily primed PCR for epidemiologic analysis of methicillin-resistant *Staphylococcus aureus* nosocomial infections. J Clin Microbiol 36:1128–1134
16. Panlilio AL, Culver DH, Gaynes RP, Banerjee S, Henderson TS, Tolson JS, Martone WJ (1992) Methicillin-resistant *Staphylococcus aureus* in U.S. hospitals, 1975–1991. Infect Control Hosp Epidemiol 13:582–586
17. Reischl U, Pulz M, Ehret W, Wolf H (1994) PCR-based detection of mycobacteria in sputum samples using a simple and reliable DNA extraction protocol. BioTechniques 17:844–845
18. Reischl U, Linde HJ, Metz M, Leppmeier B, Lehn N (2000) Rapid identification of methicillin-resistant *Staphylococcus aureus* and simultaneous species confirmation using real-time fluorescence PCR. J Clin Microbiol 38: 2429–2433
19. Ribeiro J, Vieira FD, King T, D'Arezzo JB, Boyce JM (1999) Misclassification of susceptible strains of *Staphylococcus aureus* as methicillin-resistant *S. aureus* by a rapid automated susceptibility testing system. J Clin Microbiol 37:1619–1620
20. Ruane PJ, Morgan MA, Citron DM, Mulligan ME (1986) Failure of rapid agglutination methods to detect oxacillin-resistant *Staphylococcus aureus*. J Clin Microbiol 24:490–492
21. Schwarzkopf A, Karch H, Schmidt H, Lenz W, Heesemann J (1993) Phenotypical and genotypical characterization of epidemic clumping factor-negative, oxacillin-resistant *Staphylococcus aureus*. J Clin Microbiol 31:2281–2285
22. Smith TL, Pearson ML, Wilcox KR, Cruz C, Lancaster MV, Robinson-Dunn B, Tenover FC, Zervos MJ, Band JD, White E, Jarvis WR (1999) Emergence of vancomycin resistance in *Staphylococcus aureus*. Glycopeptide-Intermediate *Staphylococcus aureus* Working Group. N Engl J Med. 340:493–501
23. Vannuffel P, Laterre PF, Bouyer M, Gigi J, Vandercam B, Reynaert M, Gala JL (1998) Rapid and specific molecular identification of methicillin-resistant *Staphylococcus aureus* in endotracheal aspirates from mechanically ventilated patients. J Clin Microbiol 36:2366–2368
24. Voss A, Milatovic D, Wallrauch-Schwarz C, Rosdahl VT, Braveny I (1994) Methicillin-resistant *Staphylococcus aureus* in Europe. Eur J Clin Microbiol Infect Dis 13:50–55

25. Wenzel RP, Reagan DR, Bertino, JS Jr, Baron EJ, Arias K (1998) Methicillin-resistant *Staphylococcus aureus* outbreak: a consensus panel's definition and management guidelines. Am J Infect Control 26:102–110
26. Wilkerson M, McAllister S, Miller JM, Heiter BJ, Bourbeau PP (1997) Comparison of five agglutination tests for identification of *Staphylococcus aureus*. J Clin Microbiol 35:148–151

Rapid Detection of Group B Streptococci Using the LightCycler Instrument

Danbing Ke, Christian Ménard, François J. Picard, Michel G. Bergeron*

Introduction

Group B streptococci (GBS), or *Streptococcus agalactiae*, has remained the leading cause of bacterial sepsis and meningitis in neonates for the last two decades [1]. The incidence of perinatal group B streptococcal disease in the United States has been decreasing because intrapartum antibiotic prophylaxis has been widely used for prevention of GBS disease [2]. Identification of GBS-colonized women is critical for prevention of neonatal GBS infections. Currently, prenatal screening culture, including broth culture in selective medium, is the gold standard method for detection of anogenital GBS colonization [1]. However, culture methods require up to 48 h to yield results and only predict 87% of women likely to be colonized by GBS at delivery [3]. A rapid, sensitive and specific test for detection of GBS directly from clinical specimens would allow for a simpler and more efficient prevention program.

Rapid tests, such as rapid antigen-based tests, have been developed but are neither sensitive nor specific enough to substitute for bacterial culture [4]. Hybridization-based methods have been successfully used for the detection of GBS from broth cultures, but remain insufficiently sensitive for detection and identification of GBS directly from clinical specimens of colonized women [5]. GBS-specific PCR-based assays have demonstrated better sensitivity but required complicated procedures which are not applicable to clinical use [6, 7].

GBS can be presumptively identified by the CAMP (named for Christie, Atkins, Munch-Petersen) test, based on detection of a diffusible extracellular protein (CAMP factor) produced by the majority of GBS [8]. The *cfb* gene encoding the CAMP factor is present in virtually every GBS isolate and is an obvious candidate for the development of a PCR assay for identification of GBS [8]. In this study, a pair of GBS-specific PCR amplification primers was designed for amplifying a segment within the *cfb* gene and evaluated by PCR using the LightCycler instrument, which allowed for shorter running time and real-time detection of amplicons by using fluorescence measurements. The assay was shown to be specific and highly sensitive for the detection and identification of GBS directly from clinical samples.

* Michel G. Bergeron (✉) (e-mail: Michel.G.Bergeron@crchul.ulaval.ca)
Centre de Recherche en Infectiologie de l'Université Laval, 2705 Boul. Laurier, Sainte-Foy, Québec, G1V 4G2, Canada

Materials

Equipment

LightCycler instrument (Idaho Technologies, Idaho Falls, ID)
Oligo primer analysis software (vers. 5.0; National Biosciences, Plymouth, MN)
GCG program (Genetics Computer Group, Madison, WI)

Reagents

Amplification primers (Infectious Diseases Research Center of Laval University, Sainte-Foy, Québec, Canada)
Adjacent Hybridization Probes (Operon Technologies, Alameda, CA)
KlenTaq1 DNA polymerase (AB peptides, Saint Louis, MI)
TaqStart Antibody (Clontech Laboratories Inc., Palo Alto, CA)
G NOME kit (Bio101, Vista, CA)
Original TA cloning kit (Invitrogen Corp., Carlsbad, CA)
QIAGEN plasmid mini-kit (QIAGEN Inc., Mississauga, Ontario, Canada)
IDI DNA extraction kit [Infectio Diagnostic (IDI) Inc., Sainte-Foy, Québec, Canada]

Procedure

Sample Preparation

Genomic DNA from a battery of bacteria including GBS was prepared using the G NOME kit. Combined vaginal and anal specimens were processed according to the CDC recommendations for culture [1] and using the IDI DNA extraction kit for PCR according to the manufacturer's instructions.

Oligonucleotides

The *cfb* gene sequences available in GenBank [*Streptococcus agalactiae* (X72754), *Streptococcus uberis* (U34322) and *Streptococcus pyogenes* (AF079502)] were aligned with the GCG package to compare the homology of these *cfb* sequences. The *cfb* sequences from two GBS strains described by Podbielski et al. [8] were also used to identify regions conserved in GBS only. PCR primers and hybridization probes (Table 1) complementary to these conserved regions were analyzed using the Oligo primer analysis software.

Two pairs of fluorescently labeled adjacent hybridization probes (STB-F/STB-C hybridizing to GBS-specific amplicons and IC-F/IC-C hybridizing to the internal control amplicon) were synthesized and HPLC-purified by Operon Technologies (Table 1). The hybridization of both adjacent probes to their target sequence, which are separated by a single nucleotide, allow fluorescence resonance energy transfer to generate an increased fluorescence signal. The probes STB-F and IC-F were labeled with fluorescein while STB-C and IC-C were labeled with Cy5. The Cy5-labeled probes contained a 3′-blocking phosphate group to prevent extension of the probes during the PCR amplification.

Internal Control

An internal control was constructed essentially as previously described by Rosenstraus et al. [9]. A 252-bp DNA fragment consisting of a 206-bp sequence not found in GBS flanked by the sequences of each of the two GBS-specific primers was used as a template for the internal control. This fragment was cloned using the Original TA cloning kit. The recombinant plasmid was isolated from trans-

Table 1. Oligonucleotides

GBS CAMP factor gene (GenBank Accession #X72754)				
	Position	Length	GC (%)	T_m (°C)[a]
Primers				
TTTCACCAGCTGTATTAGAAGTA	369	23	34.8	57.4
GTTCCCTGAACATTATCTTTGAT	522R	23	34.8	57.4
Product	369–522	153		
Hybridization probes				
GBS-specific				
AAGCCCAGCAAATGGCTCAAA-F	447	21	47.6	60.6
Cy5-GCTTGATCAAGATAGCATTCAGTTGA	469	26	38.5	61.4
Internal control-specific				
TTATTGCAGCTTCGCCACAGGAA-F		23	47.8	62.8
Cy5-GGTCCAGCAATGTGAAGAGGCAT		23	52.2	64.6

[a]T_m=(81.5+16.6*LOG[Na+])+((41*(G+C))/length)-(500/length).

formed *E. coli* by using the QIAGEN plasmid mini kit. The purified plasmids were then linearized with *Eco*RI and serially diluted. The concentration of the linearized plasmids was optimized to permit amplification of the 252-bp internal control product without significant detrimental effect on the GBS-specific amplification.

LightCycler PCR

Strict precautions to prevent carryover of amplified DNA were used [10]. Pre- and post-PCR manipulations were conducted in separate areas. Aerosol-resistant pipette tips were used to handle all reagents and samples. Control reactions to which no DNA was added were routinely performed to verify the absence of DNA carryover.

The following master mix was used for amplification and hybridization probe-based detection of GBS:

	Volume [µl]	[Final]
KlenTaq1 buffer PC2 10×	1	1×
Primers (10 µM each)	0.4+0.4	0.4 µM (each)
GBS-specific probes (10 µM each)	0.2+0.2	0.2 µM (each)
Internal control-specific probes (10 µM each)	0.2+0.2	0.2 µM (each)
Bovine serum albumin (10 µg/µl)	0.3	0.3+0.15 µg/µl
KlenTaq1 DNA polymerase	0.02	0.5 U
TaqStart antibody	0.04	
dNTP (4 mm)	0.5	0.2 mM
Internal control template (100 copies/µl)	1	10 copies/µl
H_2O (PCR grade)	4.54	–
Total volume	9	–

To complete the amplification mixtures, 1 µl of purified DNA preparation or processed clinical sample was added to 9 µl of master mix in a 0.2-ml tube and mixed by vortexing. Subsequently, 7 µl of the mixture was transferred into each glass capillary. After a short centrifugation, the capped capillaries were placed into the LightCycler sample carousel.

Concomitant amplification of the internal control allowed verification of the efficiency of the PCR to ensure that there was no significant PCR inhibition by the test sample. However, the internal control was amplified in a separate reaction vessel since in the LightCycler model used, only two fluorescent signals can be monitored in the same capillary.

The following PCR protocol was used for amplification and hybridization probe-based detection of the GBS-specific amplicons:

- Denaturation for 3 min at 94°C
- Amplification

Parameter	Value		
Cycles	45		
Type	Quantification		
	Segment 1	Segment 2	Segment 3
Target temperature [°C]	95	55	72
Incubation time [s]	0	14	5
Temperature transition rate [°C/s]	20	20	20
Acquisition mode	None	Single	None
Gains	F1=4; F2=16		

Results

Specificity

The specificity of the real-time PCR assay was verified by testing genomic DNA from bacterial species that are phylogenetically close to GBS, including members of the genera *Streptococcus*, *Lactococcus*, *Staphylococcus*, *Enterococcus*, *Abiotrophia*, *Peptostreptococcus*, and *Listeria*. Some species encountered in the normal vaginal and anal flora, such as *Bifidobacterium breve*, *Clostridium difficile*, and *Bacterioides fragilis*, were also tested. Only genomic DNA from GBS was amplified [11].

Sensitivity

The analytical sensitivity (i.e., minimal number of genome copies that can be detected) of the PCR assays was determined by using serial 10-fold and 2-fold dilutions of purified genomic DNA from 5 GBS ATCC strains. An example of the sensitivity test with the LightCycler PCR assay is given in Fig. 1. The detection limit for all five ATCC strains was one genome copy of GBS. In order to evaluate the efficiency of the IDI DNA extraction kit, three GBS-negative vaginal samples were spiked with various amounts of purified GBS genomic DNA (equivalent to 1 to 40 genome copies) and then tested by LightCycler PCR. As few as one genome copy of GBS was detected from all three GBS-negative vaginal samples spiked with genomic DNA of GBS. In terms of colony-forming units (CFUs), the PCR assay was able

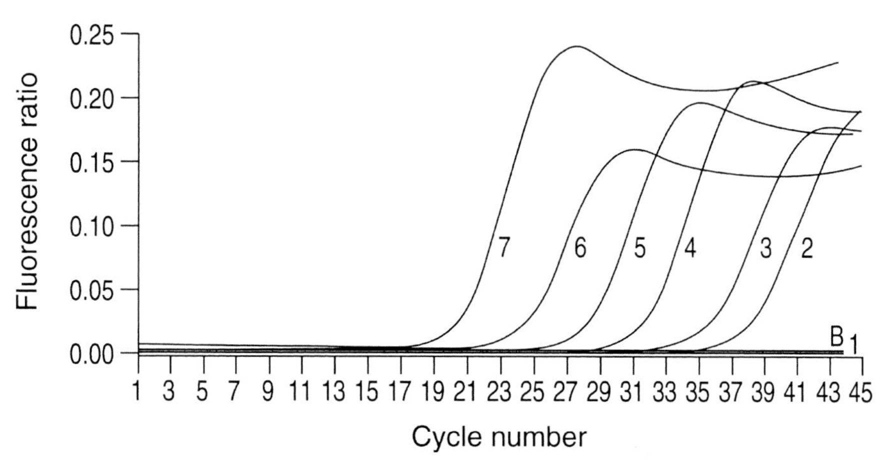

Fig. 1. Analytical sensitivity of the PCR assay for the detection of the GBS CAMP factor gene determined with serial dilutions of GBS genomic DNA. Amplification was performed from 0.5 fg (*curve 1*), 5 fg (*curve 2*), 50 fg (*curve 3*), 500 fg (*curve 4*), 5 pg (*curve 5*), 50 pg (*curve 6*), and 500 pg (*curve 7*) of genomic DNA from GBS ATCC 27591. Curve *B* represents a blank to which no DNA was added. Based on the genome size of GBS, 5 fg of GBS genomic DNA corresponds to two to three genome copies. (Adapted from [11])

to detect 1–3 CFUs from mid-log phase cultures by using the IDI DNA extraction kit as compared to 10–24 CFUs with diluted cultures added directly to the PCR mixture without pretreatment. These results confirmed the high sensitivity of our GBS-specific PCR assay as well as the efficacy of the IDI DNA extraction kit for lysis of GBS cells and prevention of significant PCR inhibition [11].

Direct Detection of GBS from Clinical Specimens

A total of 112 combined vaginal and anal specimens obtained from 112 pregnant women were tested by both the selective broth culture and the LightCycler PCR assay. The culture identified 33 GBS carriers among the 112 women tested. The LightCycler PCR assay detected GBS colonization from 32 of these 33 women. The one negative PCR result was in a sample obtained after the rupture of membranes. In the same woman, both culture and PCR were negative in the sample collected before the rupture of membranes. Using the culture as the standard, the sensitivity of the LightCycler PCR assay was 97% and the negative predictive value was 98.8%. Both the specificity and the positive predictive value were 100%. The length of time required to report results was 30–45 min for the LightCycler PCR assay and at least 36 h for culture [12]. Results of the LightCycler PCR in detection of GBS in combined vaginal and anal specimens from pregnant women are shown in Fig. 2.

Comments

Culture Methods

Currently, the gold standard method for detection of vaginal colonization with GBS is selective broth culture performed at 35–37 weeks of gestation, which is

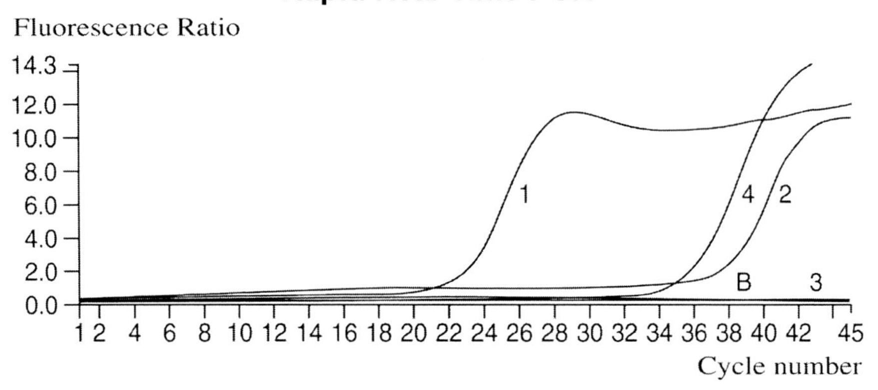

Fig. 2. Example of the LightCycler PCR assay for the detection of GBS in combined vaginal and anal specimens from pregnant women. The extent of GBS-specific amplification is measured in terms of the increase in fluorescence during the amplification process. Sample *1* was obtained from a heavily colonized woman; sample *2* was obtained from a lightly colonized woman; sample *3* was obtained from a woman with no colonization; sample *4* was a positive control, to which 10 fg of purified GBS genomic DNA had been added; and sample *B* was a negative control, to which no target DNA had been added. (Adapted from [12])

sensitive enough to allow detection of both light and heavy colonization, but identification results are not available until 36 h later [1]. Therefore, these methods are not useful for identification of GBS at or near the time of delivery. Moreover, culture results at 35–37 weeks are not always indicative of carrier status at delivery because GBS colonization is often transient [3]. Some researchers reported that overgrowth of enterococci in the selective broth media inhibited the growth of GBS [13].

LightCycler PCR Assay

In our LightCycler PCR assay, two fluorescently labeled adjacent hybridization probes complementary to part of the amplicon were used for specific hybridization to the GBS-specific amplification product. Analysis of 14 *cfb* sequences from Podbielski et al. [8] and from our own laboratory, representing the major serotypes of GBS, indicates that the target sequences for the adjacent probes are highly conserved. This ensures the ability of the two adjacent probes to hybridize to all GBS-specific amplicons. The additional specificity provided by the combination of species-specific primers with internal probes allows direct detection and identification of GBS from vaginal and anal specimens. Although all GBS strains tested were positive for the CAMP test, the assays should be able to identify CAMP-negative strains because the *cfb* gene is present in virtually all GBS isolates according to phenotypic and molecular characterizations [8].

The LightCycler platform used in this study allowed monitoring of only two fluorescence signals in each capillary. Therefore, a second capillary was needed to verify the presence of PCR inhibitors in the tested samples by using the other pair of adjacent hybridization probes to target the internal control template. A new generation of LightCycler instrumentation developed by Roche Diagnostics allows monitoring of three fluorescence signals in the same capillary, thereby permitting the use of two pairs of adjacent probes (labeled with LCRed640 or LCRed705 as acceptor flurophores) in the same reaction vessel.

Internal Control

The internal control used in this study allowed for amplification of two different target sequences with a single pair of primers. This design significantly decreases the complexity of the GBS-specific assay because few primers were used. We also optimized the amount of the internal control template used to minimize the detrimental effect of internal control amplification on GBS-specific amplification.

Sample Preparation

The IDI DNA extraction kit is simple, rapid and efficient for amplifying GBS DNA from combined vaginal and anal specimens. There was a nearly perfect correlation between the standard culture method and the LightCycler PCR assay in our clinical study [12], showing the efficiency of the IDI sample preparation protocol. Integration of rapid sample preparation with real-time PCR technologies, such as the LightCycler instrument, will help to improve the management and prevention of GBS disease as clinicians will rapidly have in hand the clinical microbiology results.

References

1. Centers for Disease Control and Prevention (1996) Prevention of perinatal group B streptococcal disease: a public health perspective (RR-27). MMWR Morb Mortal Wkly Rep 45:1–24
2. Schrag SJ, Zywicki S, Farley MM, Reingold AL, Harrison LH, Lefkowitz LB, Hadler JL, Danila R, Cieslak PR, Schuchat A (2000) Group B streptococcal disease in the era of intrapartum antibiotic prophylaxis. N Engl J Med 342:15–20
3. Yancey MK, Schuchat A, Brown LK, Lee Venture V, Markenson GR (1996) The accuracy of late antenatal screening cultures in predicting genital group B streptococcal colonization at delivery. Obstet Gynecol 88:811–815
4. Baker CJ (1996) Inadequacy of rapid immunoassays for intrapartum detection of group B streptococcal carriers. Obstet Gynecol 88:51–55
5. Bourbeau PP, Heiter BJ, Figdore M (1997) Use of Gen-Probe AccuProbe Group B streptococcus test to detect group B streptococci in broth cultures of vaginal-anorectal specimens from pregnant women: comparison with traditional culture method. J Clin Microbiol 35:144–147
6. Backman A, Lantz P-G, Rådstrom P, Olcén P (1999) Evaluation of an extended diagnostic PCR assay for detection and verification of the common causes of bacterial meningitis in CSF and other biological samples. Mol Cell Probes 13:49–60
7. Mawn JA, Simpson AJ, Heard AJ (1993) Detection of the C protein gene among group B streptococci using PCR. J Clin Pathol 46:633–636
8. Podbielski A, Blankenstein O, Lutticken R (1994) Molecular characterization of the *cfb* gene encoding group B streptococcal CAMP-factor. Med Microbiol Immunol 183:239–256
9. Rosenstraus M, Wang Z, Chang SY, DeBonville D, Spadoro JP (1998) An internal control for routine diagnostic PCR: design, properties, and effect on clinical performance. J Clin Microbiol 36:191–197
10. Neumaier M, Braun A, Wagener C (1998) Fundamentals of quality assessment of molecular amplification methods in clinical diagnostics. International Federation of Clinical Chemistry Scientific Division Committee on Molecular Biology Techniques. Clin Chem 44:12–26

11. Ke D, Ménard C, Picard FJ, Boissinot M, Ouellette M, Roy PH, Bergeron MG (2000) Development of conventional and real-time PCR assays for the rapid detection of group B streptococci. Clin Chem 46:324–331
12. Bergeron MG, Ke D, Ménard C, Picard FJ, Gagnon M, Bernier M, Ouellette M, Roy PH, Marcoux S, Fraser WD (2000) Rapid detection of group B streptococci in pregnant women at delivery (see comments). N Engl J Med 343:175–179
13. Rosa-Fraile M, Rodriguez-Granger J, Cueto-Lopez M, Sampedro A, Gaye EB, Haro JM, Andreu A (1999) Use of Granada medium to detect group B streptococcal colonization in pregnant women. J Clin Microbiol 37:2674–2677

Rapid Detection and Quantification of *Chlamydia trachomatis* in Clinical Specimens by LightCycler PCR

Heidi Wood*, Udo Reischl, Rosanna W. Peeling

Introduction

Chlamydiae are obligate intracellular bacterial pathogens. Ocular strains of *Chlamydia trachomatis* infection are the leading infectious cause of blindness with 500 million people estimated to be at risk. Genital strains of *Chlamydia trachomatis* are the most prevalent cause of bacterial sexually transmitted disease. Women with cervical chlamydial infections are at risk of developing pelvic inflammatory disease, which may lead to long-term reproductive sequelae, such as chronic pelvic pain, ectopic pregnancy, and tubal infertility.

Chlamydiae have a unique bi-phasic life cycle. The elementary bodies of chlamydiae are the infectious forms of chlamydiae. They are metabolically inert but are able to invade mammalian cells. Once inside the host cell, the elementary bodies transform into reticulate bodies, which can utilise host energy resources to fuel its own metabolism and undergo binary fission in a membrane bound inclusion. At 48–72 h post infection, when the host metabolites become limiting, the reticulate bodies transform to elementary bodies which then exit the cell to infect new hosts. The invasion of chlamydia elementary bodies into a host cell creates a stressful condition in the host as well as a hostile environment for the chlamydiae, as the host mounts an immune response against the invading microbe. Thus, an up-regulation of the heat shock response frequently occurs in both the chlamydiae and the host during an infection. Heat shock proteins (hsps) are among the most highly conserved and abundant molecules in nature. These proteins play an essential role in the folding, unfolding, and translocation of proteins, as well as in the assembly and disassembly of protein complexes. Heat shock proteins also represent dominant antigens in a wide spectrum of microbial infections [8]. Studies of human chlamydial infection have shown that antibody response to a 60-kDa chlamydial heat shock protein, Chsp60, is associated with the development of adverse sequelae following ocular and genital chlamydial infection [4, 5, 7]. However, the mechanism by which Chsp60 contributes to the immunopathological sequelae associated with chlamydial infection remains unclear. It is speculated that antibody response to Chsp60 may be a marker of per-

* Heidi Wood, Rosanna W. Peeling (✉) (e-mail: peelingr@who.int)
Department of Medical Microbiology, University of Manitoba, Basic Medical Sciences Building, 730 William Avenue, Winnipeg, Manitoba, Canada R3E0W3

sistent chlamydial infection or of an autoimmune response elicited as a result of molecular mimicry with human hsp60. In contrast to Chsp60, antibody response to a 70-kDa chlamydial heat shock protein, Chsp70, is not associated with the development of adverse sequelae following infection with *C. trachomatis* [2]. Rather, antibodies to Chsp70 are capable of neutralising chlamydiae in vitro [6].

Using an in vitro cell culture model of persistent chlamydial infection, it was shown that the levels of Chsp60 were maintained at a normal level in persistent organisms while the synthesis of the lipopolysaccharide, the 60-kDa outer membrane protein, and the major outer membrane protein were down regulated [1]. The synthesis of Chsp70, however, was not analysed in these studies. The differential expression of these key chlamydial antigens during persistent or chronic infection may play a key role in the development of immunopathological sequelae. The expression of the Chsp60 and Chsp70 may be altered in response to various environmental stresses, such as antibiotics, cytokines, and hormones. In the present study, we describe a method for the real-time quantification of these two differentially expressed chlamydial hsp genes, one of which (Chsp70) is a single-copy gene in the chlamydia genome.

Materials

Equipment

LightCycler instrument (Roche Diagnostics, Mannheim, Germany)
LightCycler software vers. 3.3 (Roche Diagnostics)
PrimerSelect sequence analysis software (DNASTAR Inc., Madison, WI, USA)
Oligo Primer analysis software (Molecular Biology Insights, Inc., Cascade, CO)

Reagents

Amplification primers (DNA Core Facility, Health Canada, Winnipeg, Canada)
Hybridization probes (TIB MOLBIOL, Berlin, Germany)

The following reagents were purchased from Roche Diagnostics:
Bacteriophage MS2 RNA
High-Pure PCR Template Purification Kit
LightCycler FastStart DNA Master SYBR Green I kit
LightCycler FastStart DNA Master Hybridization Probes

Procedure

Sample Preparation

Chlamydia trachomatis serovar A was propagated in HeLa 229 cells. Elementary bodies were purified by gradient density centrifugation and stored at –80°C. The High Pure PCR Template Preparation Kit was used to prepare purified total genomic DNA from 200 μl of pure elementary bodies according to the manufacturer's instructions. A tenfold serial dilution of the total genomic DNA was prepared as standards ranging from 10^{-1} (30 ng/μl) to 10^{-8} (3 fg/μl).

Bacteriophage MS2 RNA was added to the 10^{-4} to 10^{-8} dilutions at a final concentration of 10 ng/μl to stabilise low concentrations of DNA.

A second set of standards was also prepared for quantitation of chlamydial organisms present in samples. This set of standards consisted of genomic DNA from purified *C. trachomatis* elementary bodies which had been quantitated using electron microscopy to determine the number of particles present in each dilution.

For a comparison of the sensitivity of conventional PCR and the SYBR Green I Chsp PCR assay developed, 25 randomly selected clinical specimens (genital swabs) previously shown to be positive for *C. trachomatis* using the Roche Diagnostics AMPLICOR CT/NG Assay (Roche Diagnostic Systems, New Jersey) were used. The swabs were processed using the Roche Specimen Preparation Kit according to manufacturer's directions and 50 ul of processed sample was used for the Roche PCR assay. For quantitative PCR, the specimens were processed as described using the Roche Specimen Preparation Kit and then further purified using the High Pure PCR Template Preparation Kit. Five µl of the samples processed by the AMPLICOR Specimen Preparation Kit were used for the SYBR Green I PCR assay.

Oligonucleotides

The primer sequences listed in Table 1 were chosen from a region of each gene which was highly conserved among all *C. trachomatis* serovars but which varied considerably from the corresponding human sequences to ensure that the primers did not amplify the human hsp60 and hsp70 genes. Both primer sets were designed to amplify the hsp60 and hsp70 genes from *C. trachomatis* serovars A-K and L_2 and not from other species of chlamydiae, such as *C. pneumoniae* or *C. psittaci*. The selection of candidate sequences for LightCycler hybridization probes was aided by the Oligo software in order to obtain comparable T_m, GC contents within the range of 35%–60%, and to avoid stable secondary structures, regions of significant self-complementarity, and stretches of palindromic sequences. For the sequence-specific detection of *C. trachomatis* hsp70 amplicons, a set of hybridization probes was selected (see Table 1). Alignments of the amplified fragments showing the annealing sites of primers and hybridization probes within the corresponding *C. trachomatis* GenBank sequence and the sequence of a closely related chlamydial species are depicted in Figs. 1 and 2.

Table 1. Oligonucleotides

C. trachomatis hsp60 and *C. trachomatis* hsp70 genes			
	Length	GC (%)	T_m (°C)
hsp60 Primers:			
GATGGTGTTACCGTTGCGA	19	52.6	66.1
CCTCCACGAATTCTGTTCAC	20	50.0	64.2
hsp70 Primers:			
f: GCCATTCATCACTATCGACG	20	50.0	65.3
r: AAGATCTCTTTACAACTGCT	20	35.0	60.9
hsp70 Probes:			
GACATTGATGATGTTCTTCTAGTTGGC-F	27	40.7	58.8
LCRed 640-AATGTCCAGAATGCCTGCGGTA	22	50.0	59.6

```
                        Primer >
PCR Prod.  1     gatggtgttaccgttgcgaaagaagttgagcttgccgacaaacatgaaaatatgggcgct 60
AE001285   1577  ............................................................ 1518
AE001600   1577  ..........t..a..t......a.c.....c.aa.............c........... 1518

PCR Prod.  61    caaatggtcaaagaagtcgccagcaaaactgctgacaaagctggagacggaactacaaca 120
AE001285   1517  ............................................................ 1458
AE001600   1517  ..g.....a.........................a..c...................... 1458

PCR Prod.  121  gctactgttcttgctgaagctatctatacagaaggattacgcaatgtaacagctggagca 180
AE001285   1457  ............................................................ 1398
AE001600   1457  ..a...........a.....a.......gc.....tc..a.a.....c..t..c..t..c 1398

PCR Prod.  181   aatccaatggacctcaaacgaggtattgataaagctgttaaggttgttgttgatcaaatc 240
AE001285   1397  ............................................................ 1338
AE001600   1397  .....t........a...a.......c..c.....c..a..a............g..c.. 1338

PCR Prod.  241   aaaaaaatcagcaaacctgttcagcatcataaagaaattgctcaagttgcaacaatttct 300
AE001285   1337  .g.......................................................... 1278
AE001600   1337  ........t..t.......a..a.....c........c........a..t..t..c..a 1278

PCR Prod.  301   gctaataatgatgcagaaatcgggaatctgattgctgaagcaatggagaaagttggtaaa 360
AE001285   1277  ............................................................ 1218
AE001600   1277  ..a.........t.c........a.....t.....a.....t.....a............ 1218

PCR Prod.  361   aacggctctatcactgttgaagaagcaaaaggatttgaaaccgttttggatattgttgaa 420
AE001285   1217  ............................................................ 1158
AE001600   1217  .....a..c..t..............t.....c..c.....t...c.c..cg....a... 1158

PCR Prod.  421   ggaatgaatttcaatagaggttacctctctagctacttcgcaacaaatccagaaactcaa 480
AE001285   1157  ............................................................ 1098
AE001600   1157  ........c.....cc.t..a........c.........t.c.................. 1098

PCR Prod.  481   gaatgtgtattagaagacgctttggttctaatctacgataagaaaatttctgggatcaaa 540
AE001285   1097  ............................................................ 1038
AE001600   1097  .....c..t............c..a.................a.....c.....a..t... 1038

PCR Prod.  541   gatttccttcctgttttacaacaagttgctgaatccggccgtcctcttcttattatagca 600
AE001285   1037  ............................................................ 978
AE001600   1037  ..c........a...............a..a.....t..a..c......t.a..c..t... 978

                                                   < Primer (antisense)
                                                   cacttgtcttaagcacctcc
PCR Prod.  601  gaagacattgaaggcgaagctttagctactttggtcgtgaacagaattcgtggagg 656
AE001285   977  ........................................................ 922
AE001600   977  .....a........a...........                               952
```

Fig. 1. Alignment of the amplified fragment within the *C. trachomatis* hsp60 gene (GenBank Accession #AE001285), and the *C. pneumoniae* hsp60 gene (GenBank Accession #AE001600) for comparison. Primer annealing sites and orientation are indicated by *bold* and *underlined letters*

LightCycler PCR

Amplification and SYBR Green I Detection of the Chsp60 Gene

The following master mix was used for amplification and SYBR Green I detection of Chsp60-specific amplicons:

	Volume [µl]	[Final]
LightCycler FastStart DNA Master SYBR Green I	2	1×
hsp60 Primers (5 µM each)	2+2	0.5 µM
$MgCl_2$ (25 mM stock)	1.6	3 mM
H_2O (PCR grade)	10.4	–
Total volume	18	–

```
                      Primer f >
PCR Prod. 1      gccattcatcactatcgacgctaatggacctaaacatttggctttaactctaactcgcgc 60
M27580    969    ............................................................ 1028
AE002336  2682   ............a..t..t........t........c..a.....g...........t.. 2741

PCR Prod. 61     tcaattcgaacacctagcttcctctctcattgagcgaaccaaacaaccttgtgctcaggc 120
M27580    1029   ............................................................ 1088
AE002336  2742   .........g..t........t.................a.....g...........a.. 2801

                                                     Ctr-HP-1 >        F  Red640
PCR Prod. 121    tttaaaagatgctaaattgtccgcttctgacattgatgatgttcttctagttggcggaat 180
M27580    1089   ............................................................ 1148
AE002336  2802   ...g...........g........a.....t..c..............t.....a..... 2861

                                      < Primer r (antisense)
                  Ctr-HP-2 >          tcgtcaacatttctctagaa
PCR Prod. 181    gtccagaatgcctgcggtacaagcagttgtaaagagatctt 221
144554    1149   ......................................... 1189
8163285   2862   ...t...........a.................         2894
```

Fig. 2. Alignment of the amplified fragment within the *C. trachomatis* hsp70 gene (GenBank Accession #M27580), and the *C. muridarum* hsp70 gene (GenBank Accession #AE002336) for comparison. Primer annealing sites, orientation and the sequence of *C. trachomatis* hsp70-specific hybridization probes are indicated by *bold* and *underlined letters*

To complete the amplification mixtures, 18 µl of master mix and 2 µl of the corresponding template DNA preparation were added to each capillary. The capillaries were then centrifuged at 1,000 rpm for 5 s and placed in the LightCycler rotor. A negative control, in which 2 µl of template was replaced with H_2O, was included in each run. A second negative control, in which 2 µl of HeLa cell DNA (20 ng) was added as template, was also included to ensure that the primer set specifically amplified *C. trachomatis* sequences, and not the corresponding host cell genes.

The following LightCycler protocol was used for amplification and SYBR Green I detection of Chsp60-specific amplicons:

- Denaturation for 10 min at 95°C (to activate the FastStart *Taq* DNA polymerase)
- Amplification

Parameter	Value		
Cycles	40		
	Segment 1	Segment 2	Segment 3
Target temperature [°C]	95	65	72
Incubation time [s]	5	10	25
Temperature transition rate [°C/s]	20	20	20
Acquisition mode	None	None	Single
Gains	F1=5		

- Cooling for 30 s at 40°C

The following LightCycler program was used for the melting curve analysis of the Chsp60-specific PCR amplicons:

Parameter	Value		
Cycles	1 Segment 1	Segment 2	Segment 3
Target temperature [°C]	95	75	95
Incubation time [s]	0	45	0
Temperature transition rate [°C/s]	20	20	0.2
Acquisition mode	None	None	Cont.
Gains	F1=5		

- Cooling for 30 s at 40°C

Amplification and SYBR Green I Detection of the Chsp70 Gene

The following master mix was used for amplification and SYBR Green I detection of the Chsp70-specific amplicons:

Component	Volume [µl]	Final concentration
LightCycler FastStart DNA Master SYBR Green I	2	1×
hsp70 Primers (5 µM each)	2+2	0.5 µM
$MgCl_2$ (25 mM stock)	4.0	6 mM
H_2O (PCR grade)	8	–
Total volume	18	–

To complete the amplification mixtures, 18 µl of master mix and 2 µl of the corresponding template DNA preparation were added to each capillary. The capillaries were then centrifuged at 1000 rpm for 5 s and placed in the LightCycler rotor. A negative control, in which 2 µl of template was replaced with H_2O, was included in each run. A second negative control, in which 2 µl of HeLa cell DNA (20 ng) was added as template, was also included to ensure that the primer set specifically amplified *C. trachomatis* sequences, and not the corresponding host cell genes.

The following PCR protocol was used for amplification and SYBR Green I detection of the Chsp70-specific amplicons:

- Denaturation for 10 min at 95°C (to activate the FastStart *Taq* DNA polymerase)
- Amplification

Parameter	Value		
Cycles	40 Segment 1	Segment 2	Segment 3
Target temperature [°C]	95	65	72
Incubation time [s]	10	10	20
Temperature transition rate [°C/s]	20	20	20
Acquisition mode	None	None	Single
Gains	F1=5		

The following program was used for the melting curve analysis of the Chsp70-specific PCR amplicons:

Parameter	Value		
Cycles	1 Segment 1	Segment 2	Segment 3
Target temperature [°C]	95	75	95
Incubation time [s]	0	45	0
Temperature transition rate [°C/s]	20	20	0.2
Acquisition mode	None	None	Cont.
Gains	F1=5		

- Cooling for 30 s at 40°C

Amplification and Hybridization Probe-Based Detection of the Chsp70 Gene

The following master mix was used for amplification and hybridization probe-based detection of the Chsp70-specific amplicons (please note that an asymmetric PCR protocol was applied using primer hsp70-f at a final concentration of 0.25 µM and primer hsp70-r at a final concentration of 0.5 µM):

	Volume [µl]	[Final]
LightCycler FastStart DNA Master Hybridization Probes	2	1×
$MgCl^2$ stock solution (25 mM)	1.6	3 mM
hsp70-f Primer (10 µM each)	0.5	0.25 µM
hsp70-r Primer (10 µM each)	1	0.5 µM
Hybridization probes (2 µM each)	2+2	0.2 µM each
H_2O (PCR grade)	2.4	
Total volume	18	

To complete the amplification mixtures, 18 µl of master mix and 2 µl of the corresponding template DNA preparation were added to each capillary. The capillaries were then centrifuged at 1000 rpm for 5 s and placed in the LightCycler rotor. A negative control, in which 2 µl of template was replaced with H_2O, was included in each run. A second negative control, in which 2 µl of HeLa cell DNA (20 ng) was added as template, was also included to ensure that the two primer sets specifically amplified *C. trachomatis* sequences, and not the corresponding host cell genes.

The following PCR protocol was used for amplification and hybridization probe-based detection of the Chsp70-specific amplicons:

- Denaturation for 10 min at 95°C (to activate the FastStart *Taq* DNA polymerase)
- Amplification

Parameter	Value		
Cycles	50 (or 40) Segment 1	Segment 2	Segment 3
Target temperature [°C]	95	55	72
Incubation time [s]	10	10	20
Temperature transition rate [°C/s]	20	20	20
Acquisition mode	None	None	Single
Gains	F1=1; F2=15		

The following program was used for the hybridization probe melting curve analysis of the Chsp70-specific PCR amplicons:

Parameter	Value		
Cycles	1 Segment 1	Segment 2	Segment 3
Target temperature [°C]	95	40	95
Incubation time [s]	0	20	0
Temperature transition rate [°C/s]	20	20	0.1
Acquisition mode	None	None	Cont.
Gains	F1=1; F2=15		

- Cooling for 2 min at 40°C

Results

PCR protocols were developed for use in the LightCycler instrument to detect and quantitate *C. trachomatis* in clinical specimens for both diagnostic and research purposes. This PCR protocol successfully amplified the targeted segments of the 60-kDa and 70-kDa heat shock protein genes of *C. trachomatis*.

Detection Limit of the SYBR Green I Chsp60 and Chsp70 LightCycler Assays (Analytical Sensitivity)

SYBR Green I and hybridization probe-based detection were performed on amplicons generated from serial dilutions of genomic DNA prepared from pure elementary bodies of *C. trachomatis*. Both primer sets were tested with ten-fold differences in starting template concentration. Both sets of primers were capable of distinguishing 10^0 to 10^{-6} serial dilutions of the genomic DNA, while only the Hsp60-specific set of primers was capable of detecting the 10^{-7} dilution. The detection limit for SYBR Green I PCR using the Hsp60-specific primers was determined to be 6 fg of genomic DNA (Fig. 3), while the detection limit for the Hsp70-specific primers was 60 fg of genomic DNA (Fig. 4). The detection limit for the hybridization probe assay using the Hsp70-specific primers was determined to be 6 fg of genomic DNA (Fig. 5).

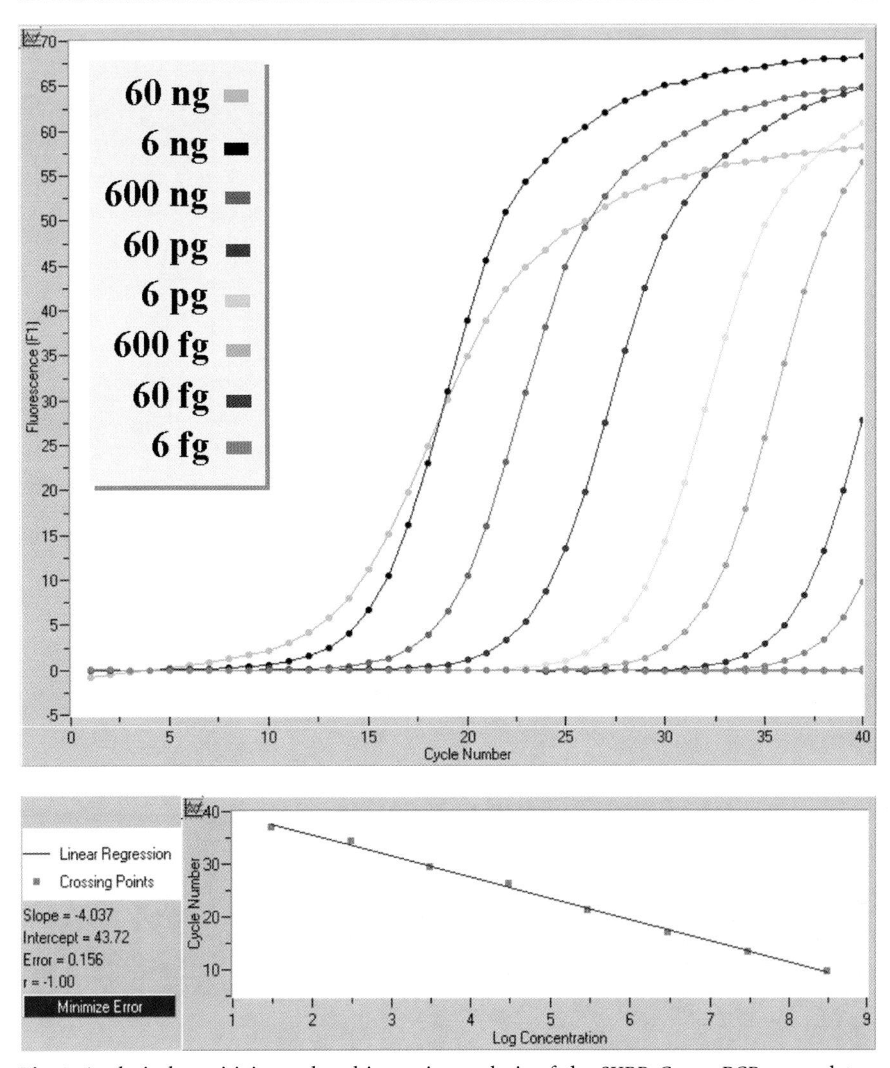

Fig. 3. Analytical sensitivity and melting point analysis of the SYBR Green PCR assay determined with serial dilutions of *C. trachomatis* genomic DNA using the hsp60-specific primers. For each dilution, the corresponding amount of *C. trachomatis* genomic DNA present in the reaction mixture is given next to the amplicon curve. Based on the slope of the original amplification curves during the log-linear phase, crossing points on the x-axis were determined by the LightCycler software. The standard curve is the linear regression line through the data points on a plot of crossing points vs logarithm of standard sample concentration. Slope, Y-intercept, mean squared error, and regression coefficient of the standard curve are given

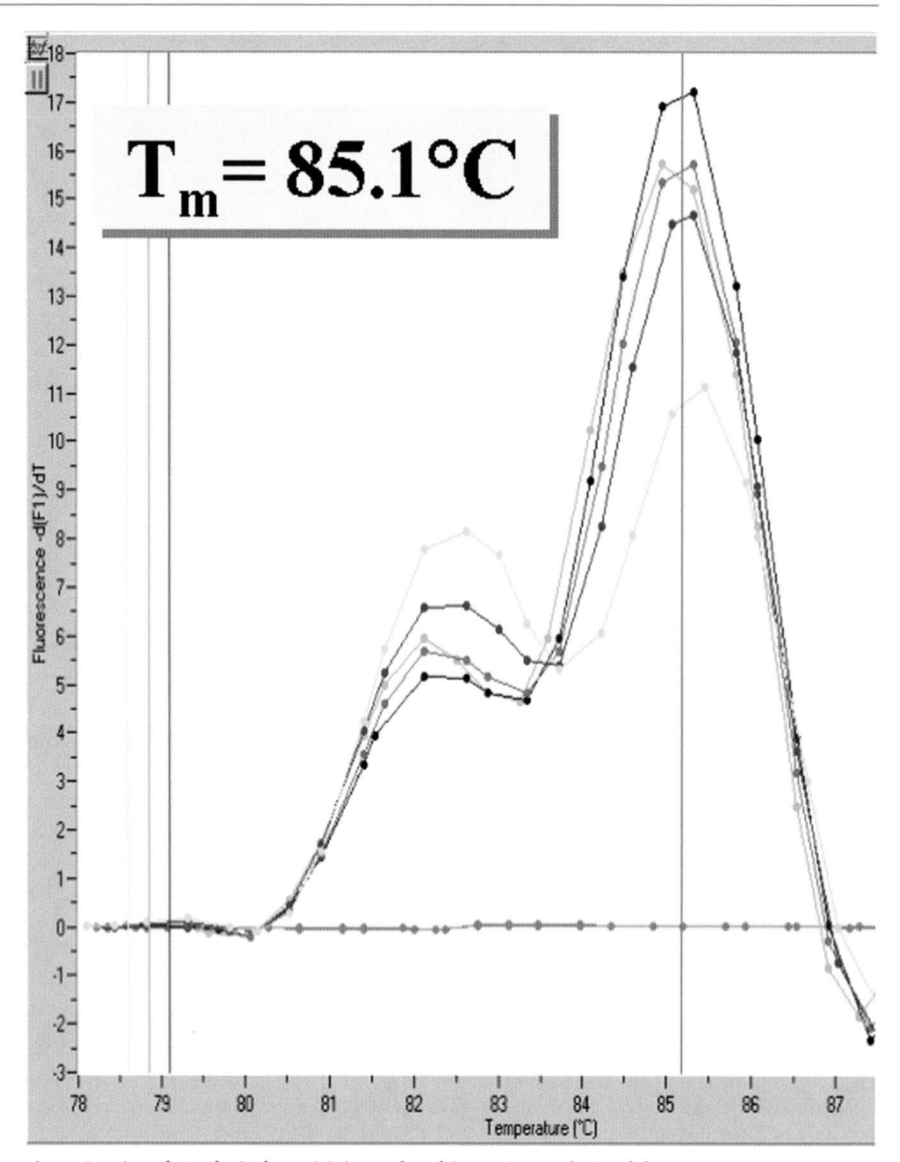

Fig. 3. *Continued.* Analytical sensitivity and melting point analysis of the SYBR Green PCR assay determined with serial dilutions of *C. trachomatis* genomic DNA using the hsp60-specific primers. For each dilution, the corresponding amount of *C. trachomatis* genomic DNA present in the reaction mixture is given next to the amplicon curve. Based on the slope of the original amplification curves during the log-linear phase, crossing points on the x-axis were determined by the LightCycler software. The standard curve is the linear regression line through the data points on a plot of crossing points vs logarithm of standard sample concentration. Slope, Y-intercept, mean squared error, and regression coefficient of the standard curve are given

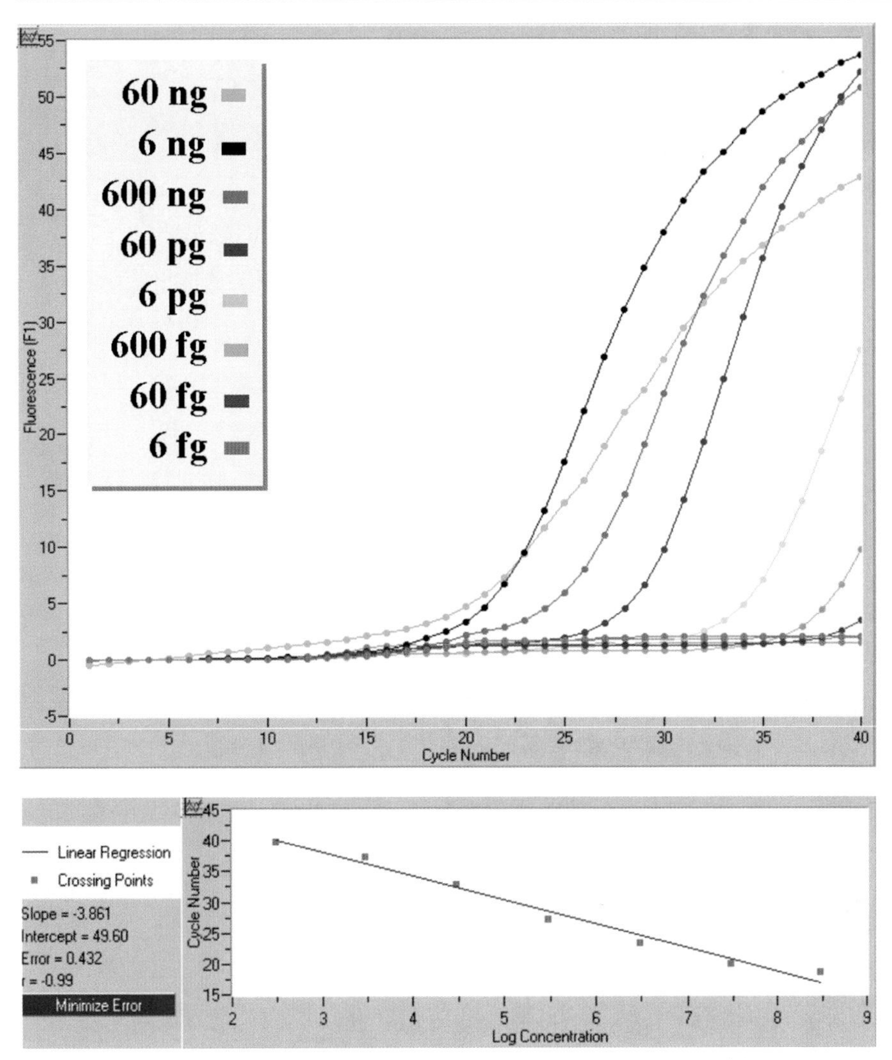

Fig. 4. Analytical sensitivity, melting point analysis and quantitative analysis of the SYBR Green PCR assay determined with serial dilutions of *C. trachomatis* genomic DNA using the Hsp70-specific primers. For each dilution, the corresponding amount of *C. trachomatis* genomic DNA present in the reaction mixture is given next to the amplicon curve

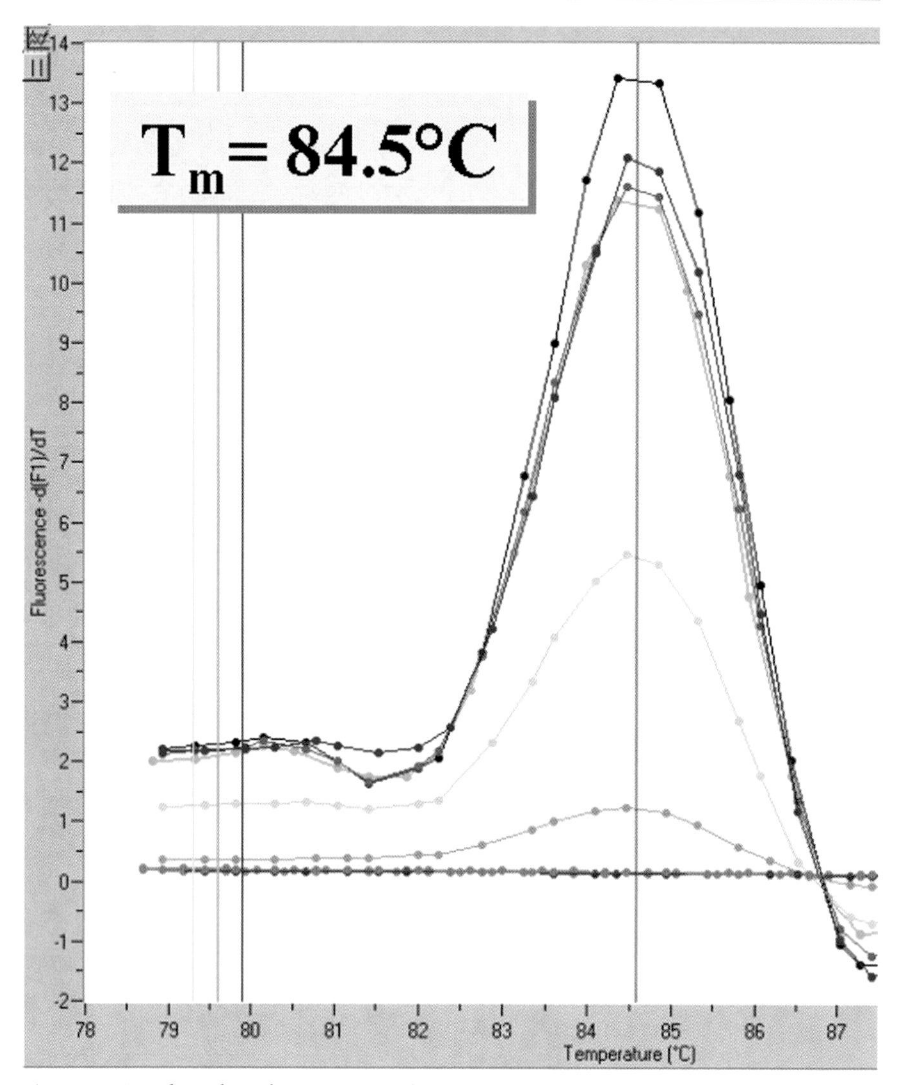

Fig. 4. *Continued.* Analytical sensitivity, melting point analysis and quantitative analysis of the SYBR Green PCR assay determined with serial dilutions of *C. trachomatis* genomic DNA using the Hsp70-specific primers. For each dilution, the corresponding amount of *C. trachomatis* genomic DNA present in the reaction mixture is given next to the amplicon curve

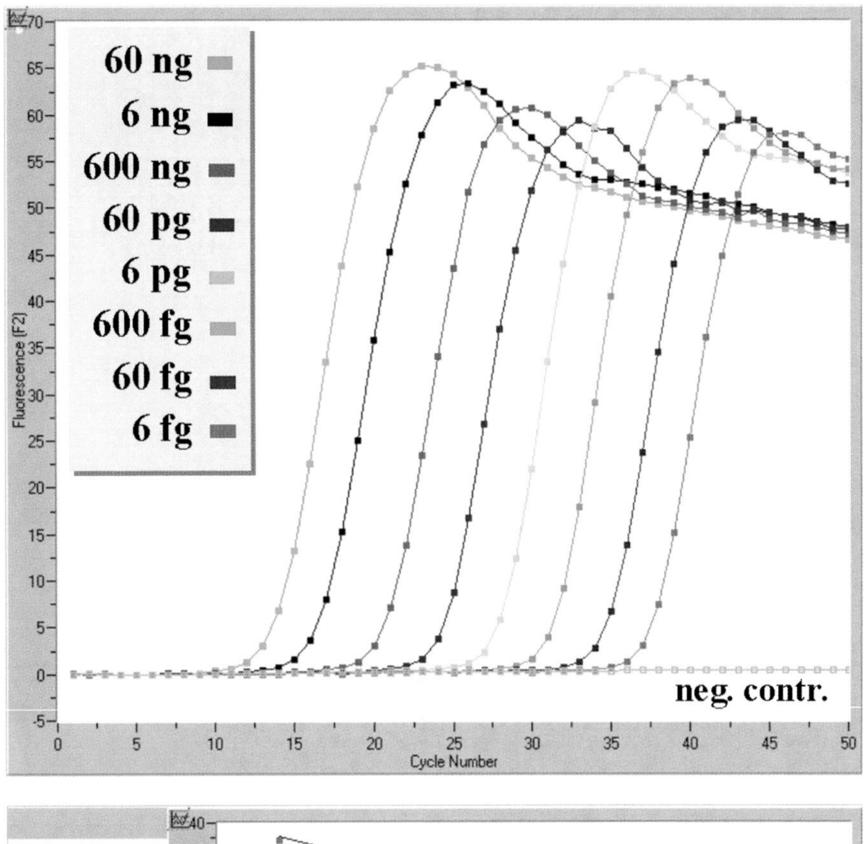

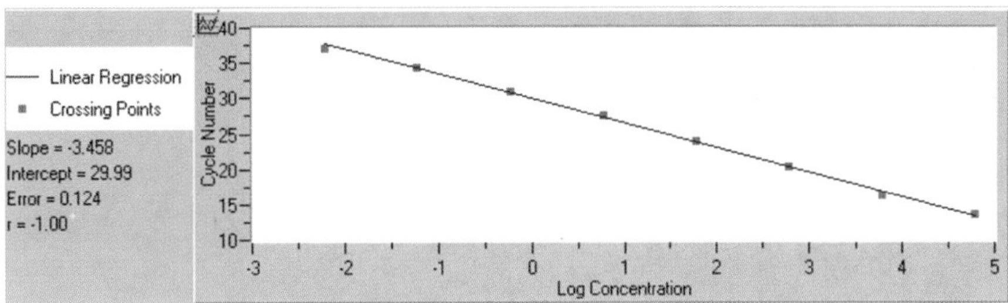

Fig. 5. Analytical sensitivity, melting point analysis and quantitative analysis of the of the Hsp70-specific hybridization probe PCR assay determined with serial dilutions of *C. trachomatis* genomic DNA. For each dilution, the corresponding amount of *C. trachomatis* genomic DNA present in the reaction mixture is given next to the amplicon curve

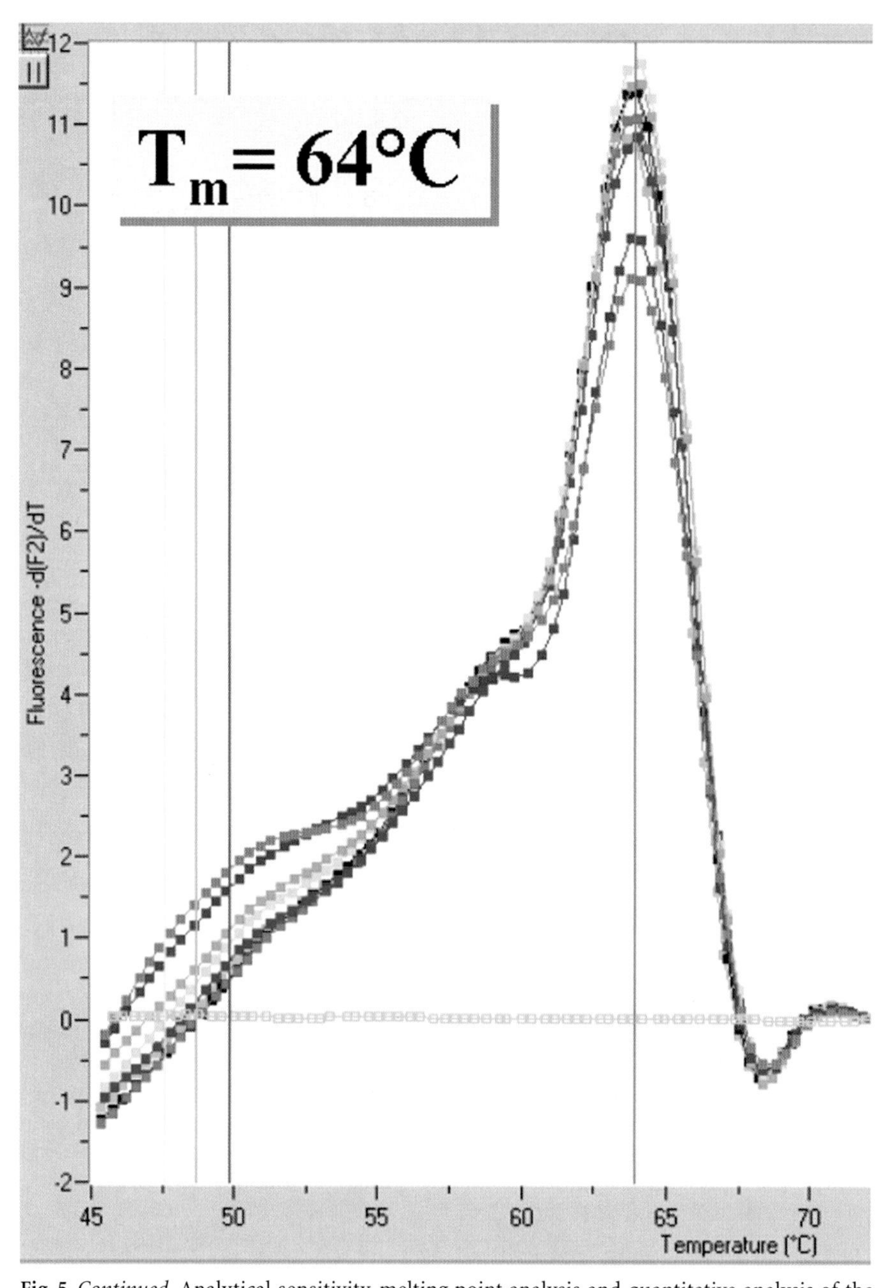

Fig. 5. *Continued.* Analytical sensitivity, melting point analysis and quantitative analysis of the of the Hsp70-specific hybridization probe PCR assay determined with serial dilutions of *C. trachomatis* genomic DNA. For each dilution, the corresponding amount of *C. trachomatis* genomic DNA present in the reaction mixture is given next to the amplicon curve

These detection limits are identical to the detection limits for conventional PCR, if the same volume (2 μl) of template DNA is used for both types of PCR. This is consistent with the finding that SYBR Green I detection of PCR products is comparable to detection by agarose gel electrophoresis.

Analytical Specificity

The PCR assays were evaluated with a collection of clinical samples originating from 25 *C. trachomatis*-positive and 50 *C. trachomatis*-negative patients. In addition, cultured strains of *C. trachomatis* serovars A-K and L_2, 2 cultured strains of *C. psittaci*, 5 strains of *C. pneumoniae*, as well as 80 isolates of gram-negative and gram-positive bacterial species other than chlamydiae were examined. LightCycler PCR results with the 75 clinical specimens were in perfect agreement with the results of previous PCR testing and all DNA preparations from cultured *C. trachomatis* serovars A-K and L_2 showed amplification by increase of fluorescence. None of the DNA preparations from the other cultured bacterial species examined were recognised (data not shown). For the hybridization probe detection format, visual examinations of the plots generated by the LightCycler software (cycle number *vs* fluorescence values of individual capillaries) allowed for a clear discrimination between positive and negative samples.

Quantitative Detection of C. trachomatis in Clinical Specimens

Of the 37 clinical specimens shown to be PCR positive in a conventional PCR assay using the Roche Amplicor assay, 28 were successfully quantitated. It was also observed that both negative controls remained negative after 40 cycles of amplification. With the use of the FastStart SYBR Green I kit, primers dimers were not observed in reactions with either the Chsp60 or the Chsp70 PCR reactions. The quantitative behaviour of the LightCycler assays is further illustrated by the perfect linearity of the regression line calculated by the LightCycler software (see Figs. 3–5). For SYBR Green I melting curve analysis, 4°C were averaged. For the quantification analysis, the Fit Points methods was used with 2 points fit.

Comments

Real-time quantitative PCR using the LightCycler instrument has proven to be an excellent alternative to conventional quantitative PCR techniques which are often based on doing serial dilutions or competitive end-point PCR. Conventional quantitative PCR procedures are time-consuming and laborious with the inherent flaw that they measure the amount of PCR product in the plateau phase of PCR, when the reaction components become limiting and amplification efficiency is variable. Hence the final product may bear little relationship to the starting target concentration in the reaction. Real-time quantitative PCR measurements are made during the exponential phase of amplification, which allows for an inherently more accurate and reproducible determination of the initial target concentration.

SYBR Green I Versus Hybridization Probe Detection Format

In this chapter, SYBR Green I PCR was successfully used to quantitate chlamydia using targets from two chlamydial heat shock protein genes. SYBR Green I quantitative PCR assays are usually used for rapid screening or target discovery as the assay protocol is easier to develop than assays using hybridization probes. The disadvantage is that it is relatively non-specific as the SYBR Green I dye can bind to any double-stranded DNA through its minor groove, including primer dimers. In an unbound state, SYBR Green I emits a low level of fluorescence. The fluorescence increases as more SYBR Green I dye binds more double-stranded DNA. However, the specificity of a SYBR Green I PCR assay can be significantly improved by determining the melting temperatures of the specific PCR product and the primer dimers. Once these values are known, the fluorescence signal can be acquired during each cycle at a temperature just below the T_m of the product and above the T_m of the primer dimers. This will ensure that primer dimers in the reaction mix melt apart before the SYBR Green I fluorescence signal is acquired at the end of each cycle. For maximum specificity, however, hybridization probes should be used in place of SYBR Green I. Hybridization probes will also not detect any primer dimers in the PCR reaction. SYBR Green I PCR has been shown to be useful in determining the number of copies of 16 S rRNA and to quantitate three sigma factors, rpoD, rpsD, and rpoN, throughout the chlamydial lifecycle [3]. In this study, SYBR Green I PCR was sensitive enough to detect low level transcripts from chlamydial infections during the early stages of infection.

Advantages of the FastStart Kits

Primer dimers are the products of nonspecific annealing and primer elongation events. During PCR, formation of primer dimers competes with formation of specific PCR product, leading to reduced amplification efficiency. The formation of primer dimers can be reduced, if not eliminated all together, in the LightCycler system. The FastStart *Taq* DNA polymerase is a modified form of thermostable recombinant *Taq* DNA polymerase. It is inactive at room temperature because of the heat-labile blocking groups on some of the amino acid residues of the enzyme. Therefore, there is no elongation during the period when primers can bind non-specifically. The modified enzyme is "activated" by removing the blocking groups at high temperature (i.e. a pre-incubation step at 95°C for a maximum of 10 min.

This has worked well for the Chsp PCR protocols described here in which no primer dimers were observed.

Quantitative Analysis of C. trachomatis in Clinical Specimens

Quantitative PCR has provided chlamydial researchers with a technique for quantitating bacterial load in clinical specimens and for studying chlamydial gene expression in host cells. The difficulty in accurate quantitation of gene expression for an obligate intracellular parasite is that gene expression must be standardised for the number of pathogen present in a host cell to ensure that the presence of increased mRNA or protein is not simply the consequence of increased number of pathogen inside the cell [3]. Enumeration of chamydiae particles for use as a standard in quantitative PCR is difficult as chlamydia has traditionally been quantitated using inclusion-forming units (IFU) in cell culture, or as elementary bodies by electron microscopy or immunofluorescence. IFUs do not give a true indication of absolute quantity of chlamydiae as IFUs only reflect the number of infectious chlamydiae. Moreover, the number of chlamydia particles in different inclusions is highly variable. Quantitation by electron microscopy or immunofluorescence is difficult as chlamydiae tend to form aggregates in solution. For an intracellular pathogen like chlamdyiae, it is therefore very important that the chlamydia used to calibrate the standard curve for quantitation be carefully prepared so that comparisons of bacterial load or analyses of gene expression are meaningful.

Current Limitations

A limitation of the LightCycler system for diagnosis of *C. trachomatis* infection is that the total reaction volume is only 20 µl. This limits the input of template DNA to 2–5 ul, making sensitivity of the Lightcycler assay one log less sensitive than conventional PCR, which allows approximately 50 ul of input sample. For *C. trachomatis*, an additional limitation in quantitative PCR sensitivity is that most conventional PCR assays, including the Roche Amplicor Chlamydia assay, uses the cryptic chlamydial plasmid as target. The copy number of this plasmid is estimated to be between 10–40 copies, giving it a much higher initial target copy number than quantitative PCR protocols developed using single copy genes. The chlamydia plasmid is, however, not a suitable target for quantitative PCR as there is tremendous variation in copy number amongst different serovars of *C. trachomatis*. The use of a nested protocol targeting a single copy gene, or concentrating the specimen at the purification step are possible means of increasing the sensitivity of quantitative PCR protocol on the LightCylcer instrument. Caution against contamination must be taken when using nested protocols.

Since the genome of *C. trachomatis* has been sequenced, the ability to perform real-time quantitative PCR or RT-PCR for different chlamydial genes using the LightCycler system will greatly increase our understanding of chlamydial growth and gene regulation. Real-time quantitative PCR assays will provide us with the ability to monitor gene expression throughout all stages of its unique life cycle. Knowledge of its developmental regulation may help to elucidate the pathogenesis of this important pathogen.

References

1. Beatty WL, Morrison RP, Byrne GI (1995) Reactivation of persistent *Chlamydia trachomatis* infection in cell culture. Infect Immun 63:199–205
2. Dieterle S, Wollenhaupt J (1996) Humoral immune response to the chlamydial heat shock proteins hsp60 and hsp70 in *Chlamydia*-associated chronic salpingitis with tubal occlusion. Hum Reprod 11:1352–56
3. Mathews SA, Kym MV, Timms P (1999) Development of a quantitative gene expression assay for *Chlamydia trachomatis* identified temporal expression of σ factors. FEBS Letters 458:354–358
4. Peeling RW, Kimani J, Plummer F, Maclean I, Cheang M, Bwayo J, Brunham RC (1997) Antibody to chlamydial hsp60 predicts an increased risk for chlamydial pelvic inflammatory disease. J Infect Dis 175:1153–8
5. Peeling RW, Bailey RL, Conway DJ, Holland MJ, Campbell AE, Jallow O, Whittle HC Mabey DCW (1998) Antibody response to the 60-kDa chlamydial heat shock protein is associated with scarring trachoma. J Infect Dis 177:256–9
6. Danilition SL, Maclean IW, Peeling RW, Winston S, Brunham RC (1990) The 75-kilodalton protein of Chlamydia trachomatis: a member of the heat shock protein 70 family? Infect Immun 58:189–96

7. Toye BC, Laferriere C, Claman P, Jessamine P, Peeling R (1993) Association between antibody to the chlamydial heat-shock protein and tubal infertility. J Infect Dis 168:1236–40
8. Zügel U, Kaufmann SHE (1999) Role of heat shock proteins in protection from and pathogenesis of infectious diseases. Clin Micro Reviews 12:19–39

Quantification of *Toxoplasma gondii* in Amniotic Fluid by Rapid Cycle Real-Time PCR

François Delhommeau, François Forestier*

Introduction

Toxoplasma gondii is the protozoan responsible for toxoplasmosis, a frequent infectious disease that may have dramatic consequences in two particular situations: immunodeficiency and pregnancy.

In immunocompromised patients such as allogeneic stem cell transplant recipients, disseminated infection and *Toxoplasma* encephalitis are usually life-threatening complications. Evaluation of *T. gondii* load by real-time quantitative PCR may be of particular interest in monitoring such patients with *Toxoplasma* infection, as it has been observed that clinical symptoms correlate with parasite number in serum and cerebrospinal fluid.

When acquired during pregnancy, acute maternal infection may result in congenital toxoplasmosis. Consequently, serious complications may occur in utero (abortion, cerebral disease), during the neonatal period, or years after birth (ocular disease). Assessment of maternal infection by serological tests leads to prenatal diagnosis, which relies on ultrasonography and amniocentesis. PCR methods are now currently used on amniotic fluid, often in association with cell cultures and mice inoculation. The use of PCR in congenital toxoplasmosis diagnosis allows accurate, rapid and sensitive answer, in contrast to also sensitive cell culture and mice inoculation, which are time consuming and give late results. Moreover, when carryover and *Taq* polymerase inhibitors are prevented, PCR methods display positive or negative results. Consequently, PCR methods help much more than fetal serology which interpretation is difficult when this method is performed.

Prognosis of congenital toxoplasmosis depends on chronological features of infection: the earlier the maternal infection is acquired during pregnancy, the lower the risk is of transmission to the fetus, but the more dramatic the consequences are. Quantification of *T. gondii* in amniotic fluid by rapid cycle real-time PCR may help in establishing new predictive criteria and in evaluating the therapeutic efficiency in case of maternal infection. In this preliminary work, our aim was to study the feasibility and the reliability of such a quantification in amniotic fluid.

* François Delhommeau, François Forestier (✉) (e-mail: ippffbio@club.internet.fr)
Laboratoire de Biologie Périnatale, Institut de Puériculture de Paris,
26 Bd Brune, 75014 Paris, France

Materials

Equipment

LightCycler instrument (Roche Diagnostics, Meylan, France)
LightCycler software vers. 3 (Roche Diagnostics)
LightCycler Carousel Centrifuge (Roche Diagnostics)

Reagents

Amplification Primers (TIB MOLBIOL, Berlin, Germany)
Hybridization Probes (TIB MOLBIOL)

The following reagents were purchased from Roche Diagnostics:
High Pure PCR Template Preparation kit
LightCycler FastStart DNA Master Hybridization Probes
Uracil-*N*-Glycosylase (UNG) for carryover prevention

Other reagents:
T. gondii genomic DNA (RH strain; kindly provided by Dr. E. Delabesse, Laboratoire de Parasitologie, Groupe Hospitalier Pitié-Salpétrière, Paris, France): concentrations were measured with a spectrophotometer and converted in equivalent tachyzoite number using the following relationship: 0.1 pg DNA=1 tachyzoite. Serial dilutions of this DNA in water were used as standards for quantification assay.

Procedure

Sample preparation

A 2-ml sample of *T. gondii* PCR positive amniotic fluid was centrifuged and the supernatant was discarded. The cell pellet was then processed for DNA extraction using the High Pure PCR Template Preparation kit, according to the manufacturer's instructions. Following the centrifugation and wash steps, total DNA was eluted with 100 µl of elution buffer and a 5 µl aliquot was directly transferred to PCR.

Oligonucleotides

Primers and hybridization probes were designed in order to amplify and hybridize to *T. gondii*- repeated B1 gene (approx. 35 copies per genome), a common target for conventional PCR detection of the parasite. Forward primer, reverse primer, fluorescein, and LCRed640 hybridization probes are described in Fig. 1 and Table 1.

Table 1. Oligonucleotides

T. gondii B1 gene (GenBank Accession #AF179871)				
	Length	GC (%)	T_m (°C)	Purity
Primers				
AAATGTGGGAATGAAAGAGACGC	23	43.5	66.4	ND
GACCAATCTGCGAATACACCAAAGT	25	44.0	68.9	ND
Probes				
GCATAGGTTGCAGTCACTGACGAGCT-F	26	53.8	73.1	ND
LCRed640-CCTCTGCTGGCGAAAAGTGAAATTCAT	27	44.4	70.8	ND

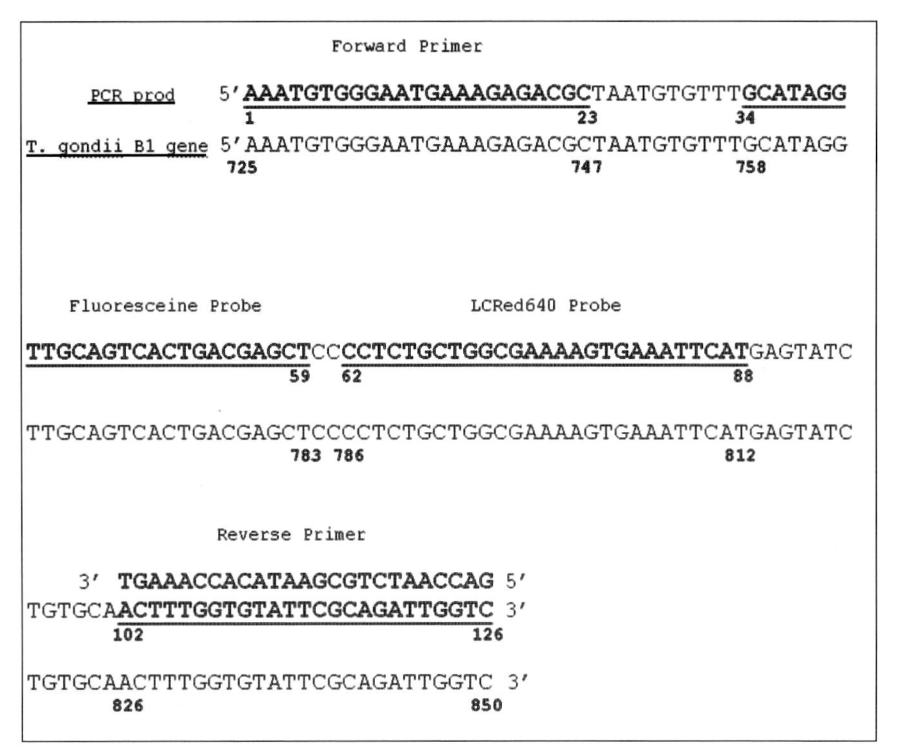

Fig 1. Sequences and positions of target gene (GenBank Accession # AF179871), PCR product, primers and hybridization probes (*bold* and *underlined*)

LightCycler PCR

LightCycler PCR was performed in a total volume of 20 µl using the LightCycler PCR master mix and 5 µl of DNA template from standard and amniotic fluid samples.

	Volume [µl]	[Final]
LightCycler FastStart DNA Master Hybridization Probes	2	1×
$MgCl_2$ stock solution	3.2	5 mM
Primers (10 µM each)	1+1	0.5 µM each
Fluorescein probe (5 µM)	1	0.25 µM
LCRed640 probe (5 µM)	1.5	0.375 µM
UNG (1 unit/µl)	0.3	0.015 unit/µl
H_2O (PCR grade)	5	
Total volume	15	

The following protocol was used:
- Denaturation for 10 min at 95°C
- Amplification

Parameter	Value		
Cycles	50		
Type	Quantification		
	Segment 1	Segment 2	Segment 3
Target temperature [°C]	95	60	72
Incubation time [s]	10	10	15
Temperature transition rate [°C/s]	20	20	2
Acquisition mode	None	Single	None
Fluorimeter gain	F1=1; F2=15; F3=30		

- Cooling for 2 min at 40°C

Results

Quantification of *T. gondii* was performed on 51 amniotic fluid samples, tested positive for *T. gondii* in conventional block cycler PCR. Figure 2 shows the amplification runs, standard curve, and the results of two representative low parasite concentration samples using second derivative method analysis. As shown in Fig. 3A, parasite concentrations ranged from 0.4 to 1096 tachyzoites/ml (median 14.8). No sample was found to be negative (sensitivity=100%). Forty one percent of amniotic fluid samples (21/51) displayed concentrations below 10 tachyzoites/ml, 55% (28/51) ranged between 10 and 100 tachyzoites/ml, and only two samples reached values above 100 tachyzoites/ml. Reproducibility was found satisfactory when processing eight duplicate samples for quantification in two separate experiments (Fig. 3B).

Comments

In this preliminary work, we show that *T. gondii* quantification in amniotic fluid by rapid-cycle real-time PCR is a fast, sensitive, reproducible, and accurate method. However, it should be noted that in contrast to other targets, such as human cytomegalovirus and the parvovirus B19 in our possession, *T. gondii* load in amniotic fluid appears to be low (96% below 100 tachyzoites/ml). Moreover, one might assume that the parasite load depends on numerous variables, including time of maternal infection, term of sample collection, maternal treatment, and fetal immunity. In contrast to quantification in immunocompromised patients' blood samples, which can be repeated and monitored in a kinetic way, quantification in amniotic fluid will remain a single analysis. Therefore, available statistical studies remain essential to identify new quantitative prognostic

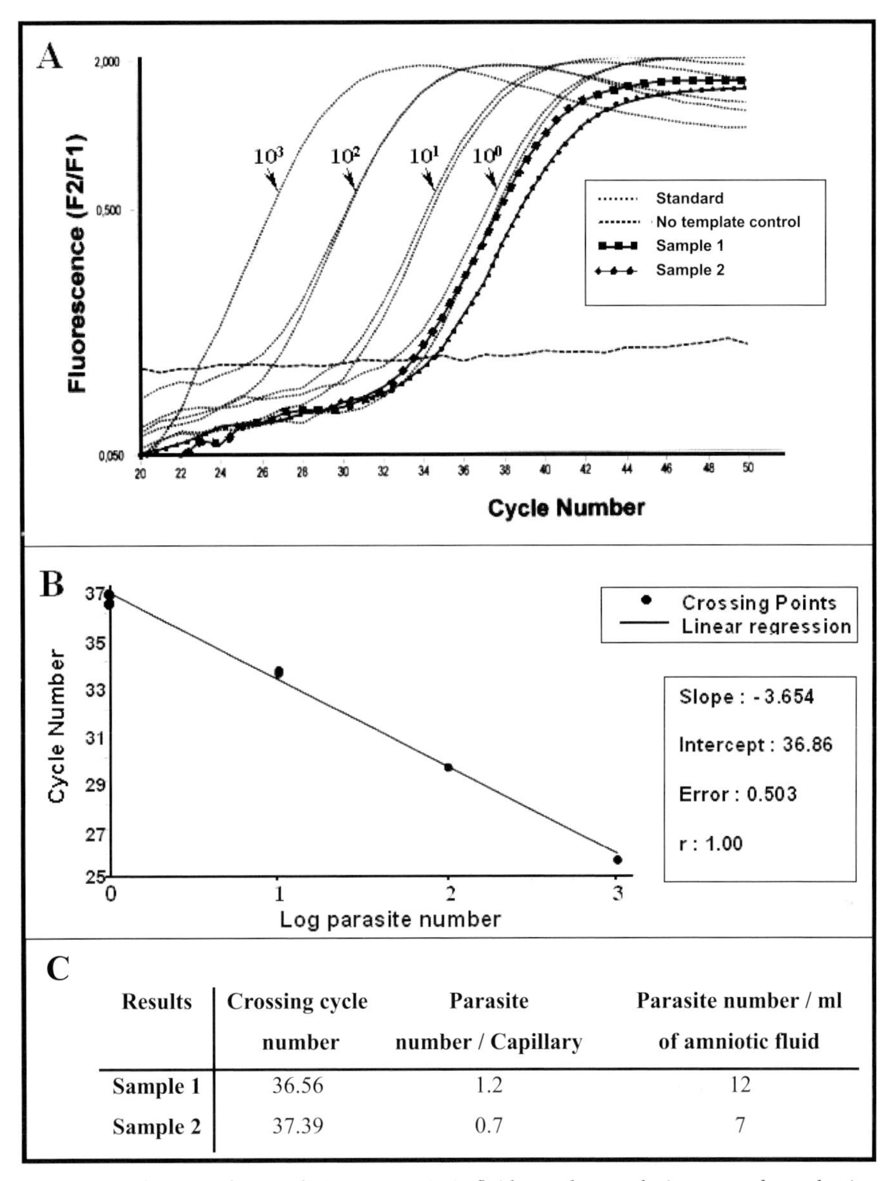

Results	Crossing cycle number	Parasite number / Capillary	Parasite number / ml of amniotic fluid
Sample 1	36.56	1.2	12
Sample 2	37.39	0.7	7

Fig 2. Quantification of *T. gondii* in two amniotic fluid samples. Analysis was performed using second derivative methods. Amplification runs of standard DNA serial dilutions (total input per capillary is indicated by *numbers above arrows*), negative control and amniotic fluid samples (A). Standard curve report (B) and Results (C)

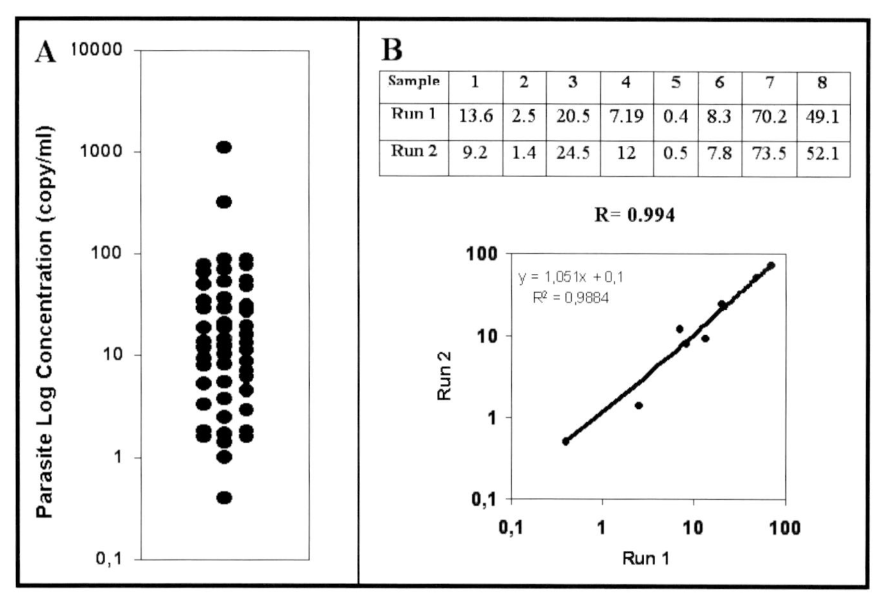

Sample	1	2	3	4	5	6	7	8
Run 1	13.6	2.5	20.5	7.19	0.4	8.3	70.2	49.1
Run 2	9.2	1.4	24.5	12	0.5	7.8	73.5	52.1

Fig 3. Quantification of *T. gondii* in amniotic fluid samples by real-time quantitative PCR. Parasite concentration in 51 amniotic fluid samples (A). Reproducibility assay performed on eight samples processed for quantification in two independent runs (run 1, x; run 2, y) (B)

and predictive criteria for congenital toxoplasmosis. Following this rationale, prospective quantification studies including a large number of amniotic fluid samples are imperative.

References

1. Costa JM, Pautas P, Ernault P, Foulet F, Cordonnier C, Bretagne S (2000) Real-time PCR for diagnosis and follow-up of toxoplasma reactivation after allogeneic stem cell transplantation using fluorescence resonance energy transfer hybridization probes. J Clin Microbiol 38:2929–2932
2. Daffos F, Forestier F, Capella-Pavlovski M, Thulliez P, Auffrant C, Valenti D, Cox WL (1988) Prenatal management of 746 pregnancies at risk for congenital toxoplasmosis. N Engl J Med 318:271–275
3. Hohlfeld P, Daffos F, Costa JM, Thulliez P, Forestier F, Vidaud M (1994) Prenatal diagnosis of congenital toxoplasmosis with a polymerase-chain-reaction test on amniotic fluid. N Engl J Med 331:695–699
4. Forestier F, Hohlfeld P, Sole Y, Daffos F (1998) Prenatal diagnosis of congenital toxoplasmosis by PCR: extended experience. Prenat Diag 18:407–409

Genospecies-Specific Melting Temperature of the *recA* PCR Product for the Detection of *Borrelia burgdorferi* sensu lato and Differentiation of *Borrelia garinii* from *Borrelia afzelii* and *Borrelia burgdorferi* sensu stricto

Johanna Mäkinen, Qiushui He, Matti K. Viljanen*

Introduction

Lyme borreliosis, manifesting as arthritis, was first reported in the USA in 1977 [1] and the expanded clinical spectrum of this tick-borne multisystem *Borrelia* infection has since been recognized in many countries. The initial symptom and hallmark of lyme borreliosis is an expanding skin rash called erythema migrans emerging around the tick-bite site. *Borrelia burgdorferi* sensu stricto, *B. garinii*, and *B. afzelii* are the genospecies responsible for human Lyme borreliosis [2]. In Europe *B. garinii* and *B. afzelii* are the most prevalent genospecies, whereas *B. burgdorferi* sensu stricto is the only genospecies encountered in North America [2].

Lyme borreliosis spirochetes are difficult to detect by culture in human samples and serologic tests are therefore the most practical and readily available methods for confirming the infection. However, the drawbacks of serology are poor sensitivity at the early stages of the disease, problems with specificity and lack of standardization. *B. burgdorferi* can be detected directly in tissue sections using immunostaining with monoclonal antibodies. This method also suffers from low sensitivity and is reliable only in experienced hands. For the detection and identification of *B. burgdorferi* sensu lato genospecies, PCR and PCR-based assays, such as species-specific PCR [3, 4], randomly amplified polymorphic DNA analysis [2, 5], PCR-based sequencing [6–10], and restriction fragment-length polymorphism [11–13] are commonly used. Although all of these methods have the advantage of high discriminative power, they are relatively laborious and expensive. However, the ability of PCR to detect *B. burgdorferi*-specific DNA in a variety of clinical specimens from patients with both early and chronic manifestations of Lyme borreliosis has made it a promising method for direct detection of *Borrelia* in clinical specimens. A biopsy from erythema migrans skin lesion is the optimal specimen for PCR. Also other biopsies and body fluids (CSF, joint fluid) can serve as specimens for PCR.

We describe a rapid and sensitive real-time PCR assay that can be used to detect *Borrelia burgdorferi* sensu lato directly in clinical specimens and further to differentiate *B. garinii* from *B. afzelii* and *B. burgdorferi* sensu stricto. The ampli-

* Matti K. Viljanen (✉) (e-mail: matti.viljanen@ktl.fi)
National Public Health Institute, Department in Turku,
Kiinamyllynkatu 13, 20520 Turku, Finland

fication target was *recA*, a chromosomal gene that belongs to a set of genes responsible for the homologous recombination in bacteria [14]. The melting curves of PCR products derived from the different genospecies of *B. burgdorferi* sensu lato, using primers described by Morrison et al. [15], were analyzed. The T_m of *B. garinii recA* PCR product was found to be about 2°C lower than that of *B. burgdorferi* sensu stricto and *B. afzelii* and, thus, it can be used to differentiate *B. garinii* from the other two genospecies.

Materials

Equipment

LightCycler instrument (Roche Diagnostics, Mannheim, Germany)
LightCycler software vers. 3.39 (Roche Diagnostics)
LightCycler Capillaries (Roche Diagnostics)
LightCycler Centrifuge Adapters (Roche Diagnostics)
LightCycler Cooling Block (Roche Diagnostics)
GeneQuant RNA/DNA calculator (Pharmacia LKB Biochrom Ltd, Cambridge, UK)

Reagents

Amplification Primers (Eurogentech, Seraing, Belgium)

The following reagents were purchased from Roche Diagnostics:
LightCycler-DNA Master SYBR Green I
Multipurpose Agarose
Bovine Serum Albumin (BSA; Roche Diagnostics)

Other reagents:
InstaGene matrix (Bio-Rad, Hercules, CA, USA)
Taq Start Antibody (Clontech, Palo Alto, CA, USA)
Human skin biopsy samples
Ixodes ricinus tick samples

Procedure

Sample Preparation

The human skin biopsy specimens were punched from the marginal areas of erythema migrans lesions. The specimens were cut in half by scissors. One half was cultured and from the other half DNA was extracted directly and used for PCR. The midgut was removed from the *Ixodes ricinus* ticks using tweezers under a microscope and cultured. DNA was extracted directly from the ticks from which the midgut was removed.

Cultivation of Bacteria

The reference strains representing *B. burgdorferi* sensu stricto, *B. garinii*, *B. afzelii*, tick midguts, and human skin biopsy specimens were inoculated into tubes containing Barbour-Stoenner-Kelly (BSK-II) medium and incubated at 30°C for 4 to 6 weeks. The tubes were inspected macroscopically twice a week. Dark-field microscopy was carried out if the color of the culture medium indi-

Table 1. Oligonucleotides

Borrelia burgdorferi s.s. Rec A gene (GenBank Accession # U23457)					
	Position	Length	GC (%)	T_m	Purity
Primer:					
GTG GAT CTA TTG TAT TAG ATG AGG CTC TCG	194	30	43.3	56.3	1.66
GCC AAA GTT CTG CAA CAT TAA CAC CTA AAG	415R	30	40	55.2	1.80
Product:	415–194	222			

cated growth. The final identification of cultured spirochetes was based on PCR and sequencing of amplicons as described earlier [16, 17].

DNA Isolation

DNA isolation from the pelleted bacterial cultures, tick midguts, and human skin biopsies was done using the InstaGene matrix according to the manufacturer's protocol. The DNA concentrations of the DNA solutions extracted from the reference strains were measured using GeneQuant RNA/DNA calculator and was adjusted to approx. 2 μg/ml.

LightCycler PCR

The following master mix was used for amplification and SYBR Green I -based detection of the the *B. burgdorferi* sensu lato species-specific amplicons:

	Volume [μl]	[Final]
LightCycler-DNA Master SYBR Green I	2	1×
$MgCl_2$ stock solution	1.6	3 mM
Primers (20μM each)	0.4 + 0.4	0.4 μM
Taq Start antibody	1	0.07 μM
BSA (5 mg/ml)	0.4	0.1 mg/ml
H_2O (PCR grade)	10.2	
Total volume	16	

Depending on the number (n) of PCR reactions, the (n+1)-fold amount of the reaction master mix was prepared and the mixture was gently vortexed. A 4-μl volume of sample DNA was added to 16 μl of the LightCycler master mix in the LightCycler glass capillaries. The capillaries were closed, centrifuged in a microcentrifuge using the LightCycler centrifuge adapters and placed in the LightCycler sample carousel.

One negative and three positive controls were included in each run. The negative control contained all the reagents except sample DNA. The three positive controls contained 4 μl of *B. afzelii*, *B. burgdorferi* sensu stricto, and *B. garinii* DNA (approx. 2 μg/ml).

The following protocol was used to detect and identify the *B. burgdorferi* sensu lato species by amplifying a segment of the *recA* gene:

- Denaturation for 40 s at 95°C
- Amplification

Parameter	Value		
Cycles	50		
Type	Quantification		
	Segment 1	Segment 2	Segment 3
Target temperature [°C]	95	59	72
Incubation time [s]	1	5	11
Temperature transition rate [°C/s]	20	20	20
Acquisition mode	None	None	Single

- Melting Curve Analysis

Parameter	Value		
Cycles	1		
Type	Melting curve analysis		
	Segment 1	Segment 2	Segment 3
Target temperature [°C]	95	55	94
Incubation time [s]	0	20	0
Temperature transition rate [°C/s]	20	20	0.1
Acquisition mode	None	None	Cont.

- Cooling for 2 min at 40°C

Gel Electrophoresis of PCR Products

The LightCycler PCR products were separated by gel electrophoresis to verify the presence of specific products. The amplified *recA* PCR products were removed from the capillary by reverse centrifugation into a 1.5-ml Eppendorf tube. A 5-μl volume of gel loading buffer was added and the samples were electrophoretically analyzed on a 1.5% agarose gel.

Results

Sensitivity

The analytical sensitivity of the assay was determined to be less than 8 fg (about five organisms, see Fig. 1) by using serial dilutions of *Borrelia* genomic DNA isolated from cultured *B. burgdorferi* sensu lato organisms. This sensitivity was comparable to that of the conventional nested PCR, targeted to the flagellin gene, which has been routinely used in our laboratory [16, 17].

Specificity

The specificity of the assay was evaluated using a collection of 32 samples containing *B. burgdorferi* sensu lato (as determined by culture and/or flagellin PCR

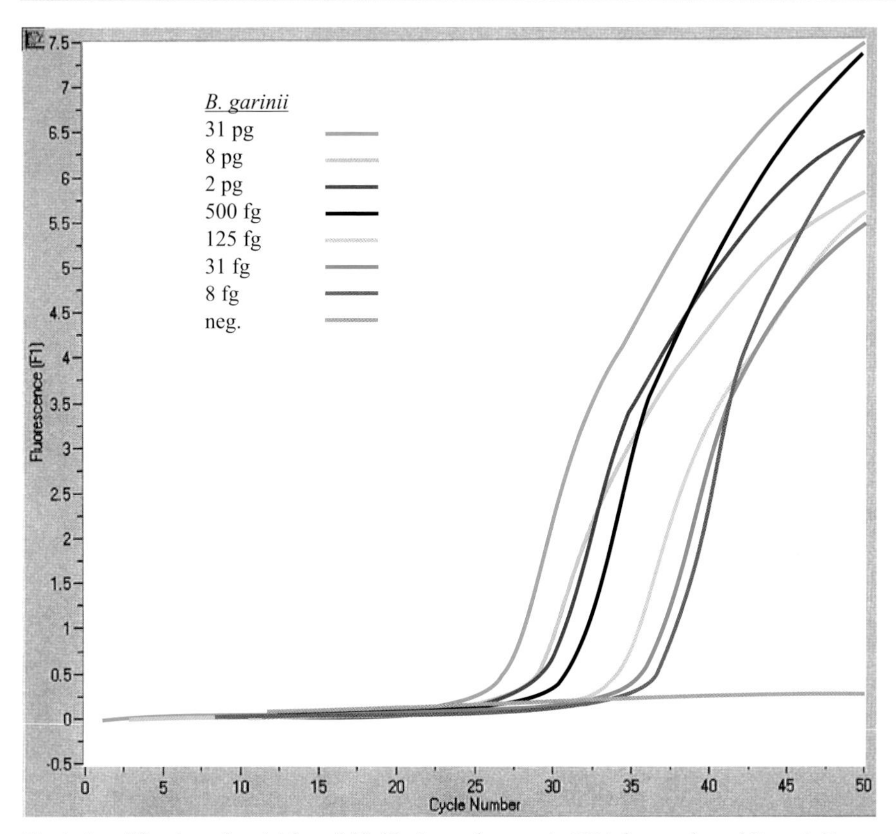

Fig. 1. Amplification of serial four-fold dilutions of genomic DNA from cultured *B. garinii*

and sequencing) originating from various sample materials, as well as cultured *B. hermsii, B. parkerii, B. turicatae*, and ten bacterial species other than *Borrelia*. All of the 32 preparations that contained *B. burgdorferi* sensu lato DNA were positive by LightCycler PCR, whereas all the other DNA preparations remained negative.

Melting Temperature

The presence or absence of specific PCR products was determined by the melting curve analysis. The mean T_m values for the reference strains *Borrelia* were *B. burgdorferi* sensu stricto, *B. afzelii*, and *B. garinii* 84.02, 83.71, and 81.62 °C, respectively (see Table 2 and Fig. 2). Unspecific products and primer dimers had a melting temperature below 80°C.

Intra-assay Variation

The intra-assay variation was studied by amplifying the DNA preparations derived from the reference strains in five different capillaries in the same run. The intra-assay coefficients of variation of the T_m for *B. burgdorferi* sensu stricto, *B. afzelii*, and *B. garinii* were 0.08, 0.05, and 0.06%, respectively.

Table 2. Origins, numbers (n), and melting temperature (T_m) of *Borrelia burgdorferi* sensu lato strains used in this study

Species/origin (n)	T_m (°C; mean±SD)
B. burgdorferi sensu stricto	
ATCC 35210 (1)	84.02
B. garinii	
387 CSF (1)	81.62
Human samples (8)	80.62±0.33
Tick samples (10)	81.30±0.40
B. afzelii	
Bo 23 (1)	83.71
Human samples (1)	83.84
Tick samples (10)	83.42±0.27

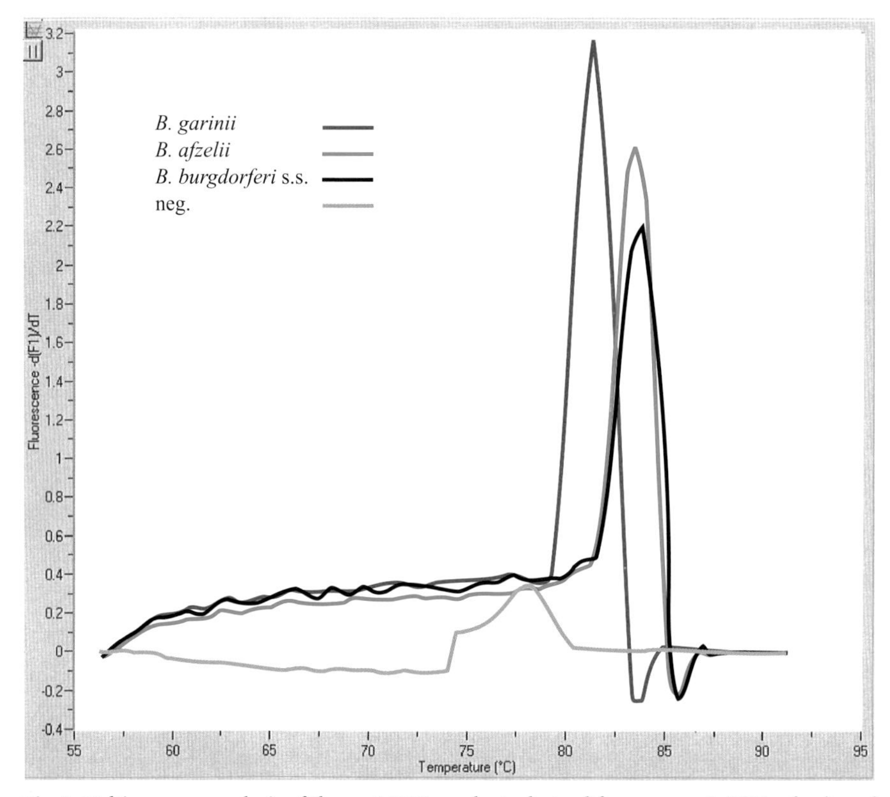

Fig. 2. Melting curve analysis of the *recA* PCR products derived from genomic DNA of cultured *B. burgdorferi* sensu lato, showing the genospecies-specific melting peaks. No specific amplification product was obtained from the negative control

Inter-assay Variation

The inter-assay variation was studied by analyzing the T_m of the three reference strains analyzed in six separate runs. The interassay coefficients of variation of the T_m for *B. burgdorferi* sensu stricto, *B. afzelii*, and *B. garinii* were 0.4, 0.4, and 0.5%, respectively.

The Effect of Template Concentration on T_m

It is known that the T_m of dsDNA in any given reaction condition is dependent on the DNA concentration [18]. The effect of template DNA concentration on the T_m was tested by amplifying six fourfold dilutions of *B. garinii* DNA (starting concentration approx. 2 µg/ml) in the same run. The mean T_m of these six different target DNA concentrations was 81.77°C, with a standard deviation of 0.41, indicating that the assay reaches a nearly equal endpoint DNA concentration, as after 50 cycles of amplification the PCR has reached the plateau phase (see Fig. 3).

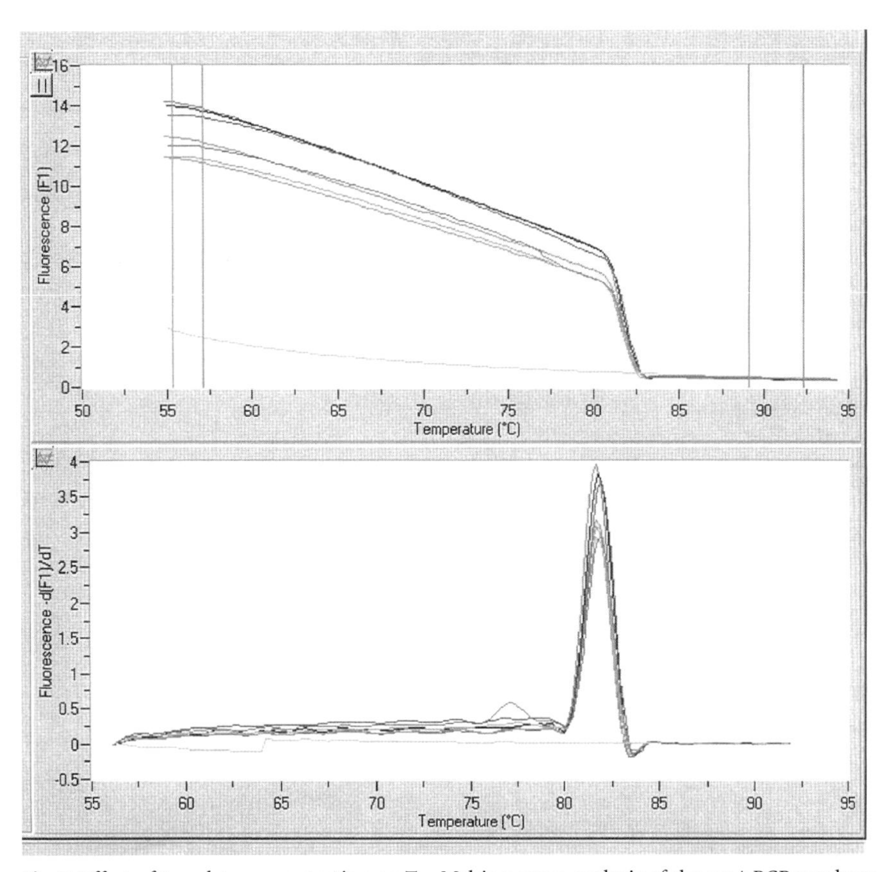

Fig. 3. Effect of template concentration on T_m. Melting curve analysis of the *recA* PCR products from serial four-fold dilutions of genomic DNA from cultured *B. garinii*, showing that the template concentration does not affect the T_m in the given reaction conditions (**A**). Melting peaks plotted as the negative derivative of fluorescence versus temperature [–d(F1)/dT] (**B**)

Comments

LightCycler PCR

The intra- and inter-assay variation coefficients of the T_m values were very low, showing that the technical principles of LightCycler allow consistent temperature conditions for the reaction capillaries. During PCR amplification, unspecific products, even primer dimers that are produced when little or no template is present, can be formed. Because these products are double-stranded, they also bind SYBR Green I dye and give a fluorescence signal. The analysis of melting curves can differentiate the unspecific products from the specific product, because the T_m of unspecific products is lower. In our experiments, all unspecific products melted below 80°C. The method was rapid and less laborious than the conventional nested PCR and PCR-based sequencing that we have previously used to detect and identify *Borrelia* organisms in various specimens. The sensitivity of this assay was also comparable to the nested PCR, but the contamination risk is much lower than in the nested procedure. Thus this assay could be used for laboratory diagnosis of human Lyme disease.

Melting Temperature

The mean T_m of *B. garinii recA* PCR products was about 2°C lower than those of the PCR products of *B. burgdorferi* sensu stricto and *B. afzelii*. A difference of 2°C or even less has been considered to be large enough for reliable differentiation of PCR amplicons [19]. The difference in T_m between *B. burgdorferi* sensu stricto and *B. afzelii* reference strains was only marginal and it cannot be used for distinguishing these two species. However, based on our preliminary sequencing data of *recA* gene, these two species could be differentiated using hybridization probes designed complementary to *B. burgdorferi* sensu stricto and/or *B. afzelii*.

Applications

It is known that the specific T_m of the amplified PCR product in given reaction conditions is a function of the GC/AT ratio, length, and nucleotide sequence. As such, T_m is a fundamental characteristic of DNA. As described in this study, it is possible to apply the melting curve analysis for the identification of PCR products, provided that there is sufficient sequence variation within the amplified region.

References

1. Steere AC, Malawista SE, Snydman DR, et al. (1977) Lyme arthritis: an epidemic of oligoarticular arthritis in children and adults in three Connecticut communities. Arthritis Rheum 20:7–17
2. Wang G, van Dam AP, Schwartz I, Dankert J (1999)Molecular typing of Borrelia burgdorferi sensu lato: taxonomic, epidemiological, and clinical implications. Clin Microbiol Rev 12:633–653
3. Marconi RT, Garon CF (1992) Development of polymerase chain reaction primer sets for diagnosis of Lyme disease and for species-specific identification of Lyme disease isolates by 16 S rRNA signature nucleotide analysis (published erratum appears in J Clin Microbiol 31:1026). J Clin Microbiol 30:2830–2834

4. Liebisch G, Sohns B, Bautsch W (1998) Detection and typing of *Borrelia burgdorferi* sensu lato in *Ixodes ricinus* ticks attached to human skin by PCR. J Clin Microbiol 36:3355–3358
5. Welsh J, Pretzman C, Postic D, Saint Girons I, Baranton G, McClelland M (1992) Genomic fingerprinting by arbitrarily primed polymerase chain reaction resolves *Borrelia burgdorferi* into three distinct phyletic groups. Int J Syst Bacteriol 42:370–377
6. Valsangiacomo C, Balmelli T, Piffaretti JC (1997) A phylogenetic analysis of *Borrelia burgdorferi* sensu lato based on sequence information from the *hbb* gene, coding for a histone-like protein. Int J Syst Bacteriol 47:1–10
7. Fukunaga M, Hamase A (1995) Outer surface protein C gene sequence analysis of *Borrelia burgdorferi* sensu lato isolates from Japan. J Clin Microbiol 33:2415–2420
8. Fukunaga M, Koreki Y (1996) A phylogenetic analysis of *Borrelia burgdorferi* sensu lato isolates associated with Lyme disease in Japan by flagellin gene sequence determination. Int J Syst Bacteriol 46:416–421
9. Gassmann GS, Jacobs E, Deutzmann R, Gobel UB (1991) Analysis of the *Borrelia burgdorferi GeHo fla* gene and antigenic characterization of its gene product. J Bacteriol 173:1452–1459
10. Misonne MC, Hoet PP (1998) Species-specific plasmid sequences for PCR identification of the three species of *Borrelia burgdorferi* sensu lato involved in Lyme disease. J Clin Microbiol 36:269–272
11. Liveris D, Wormser GP, Nowakowski J et al. (1996) Molecular typing of *Borrelia burgdorferi* from Lyme disease patients by PCR-restriction fragment length polymorphism analysis. J Clin Microbiol 34:1306–1309
12. Postic D, Assous MV, Grimont PA, Baranton G (1994) Diversity of *Borrelia burgdorferi* sensu lato evidenced by restriction fragment length polymorphism of rrf (5 S)-rrl (23 S) intergenic spacer amplicons. Int J Syst Bacteriol 44:743–752
13. Rijpkema SG, Herbes RG, Verbeek-De Kruif N, Schellekens JF (1996) Detection of four species of *Borrelia burgdorferi* sensu lato in *Ixodes ricinus* ticks collected from roe deer (*Capreolus capreolus*) in The Netherlands. Epidemiol Infect 117:563–566
14. Fraser CM, Casjens S, Huang WM et al. (1997) Genomic sequence of a Lyme disease spirochete, *Borrelia burgdorferi*. Nature 390:580–586
15. Morrison TB, Ma Y, Weis JH, Weis JJ (1999) Rapid and sensitive quantification of *Borrelia burgdorferi*-infected mouse tissues by continuous fluorescent monitoring of PCR. J Clin Microbiol 37:987–992
16. Schmidt B, Muellegger RR, Stockenhuber C et al. (1996) Detection of *Borrelia burgdorferi*-specific DNA in urine specimens from patients with erythema migrans before and after antibiotic therapy. J Clin Microbiol 34:1359–1363
17. Junttila J, Peltomaa M, Soini H, Marjamaki M, Viljanen MK (1999) Prevalence of *Borrelia burgdorferi* in *Ixodes ricinus* ticks in urban recreational areas of Helsinki. J Clin Microbiol 37:1361–1365
18. SantaLucia J Jr (1998) A unified view of polymer, dumbbell, and oligonucleotide DNA nearest-neighbor thermodynamics. Proc Natl Acad Sci USA 95:1460–1465
19. Ririe KM, Rasmussen RP, Wittwer CT (1997) Product differentiation by analysis of DNA melting curves during the polymerase chain reaction. Anal Biochem 245:154–160

Rapid and Specific Detection of *Coxiella burnetii* by LightCycler PCR

Markus Stemmler, Hermann Meyer*

Introduction

Coxiella burnetii, a zoonotic, obligate intracellular bacterium, is the causative agent of Q fever in humans as well as of abortions in domestic ruminants, primarily cattle, sheep, and goats. Humans typically acquire Q fever by inhaling infectious aerosols and contaminated dust generated by animals or animal products. The acute clinical disease associated with *C. burnetii* infection is usually a benign although temporarily incapacitating illness in humans. Even without treatment, the vast majority of patients recover. Chronic forms associated with granulomatous hepatitis or endocarditis are rare [1].

There is a spore-like form of the organism that is extremely resistant to environmental influences. This allows *C. burnetii* to remain infectious after years outside the host cell. *C. burnetii* has a remarkable infectivity, a single organism may initiate infection in an animal model. The distribution of *C. burnetii* is worldwide and it has been recovered from a large number of mammalian species and arthropods.

Diagnosis of Q fever is usually accomplished by serological testing, the most commonly used method being the indirect immunofluorescence assay. Serological testing should always be done for a patient with a febrile illness and negative blood cultures. *C. burnetii* is a very infectious organism. Thus, only biosafety level-three laboratories and experienced personnel should handle a contaminated specimen and cultivate this organism. Isolation in cell cultures is a sensitive tool, but it is rather time-consuming. A capture enzyme-linked immunosorbent assay (ELISA) has been described; however, given the low level of shedding and the minimum infective dose, it is not completely satisfactory for diagnostic purposes. PCR is a highly sensitive and specific detection method and primers targeting a repetitive, transposon-like element have been used to establish a diagnostic PCR for the detection of *C. burnetii* DNA in bovine milk [2, 3].

Here, we describe the development of a sensitive and specific hybridization probe-based LightCycler assay. Fifteen well-characterized strains of *C. burnetii* isolated worldwide from various sources and 15 isolates of Gram-negative or Gram-positive bacterial species other than *Coxiella* were tested.

* Hermann Meyer (✉) (e-mail: hermann.meyer@micro.vetmed.uni-muenchen.de)
Institut für Mikrobiologie, Sanitätsakademie der Bundeswehr, Neuherbergstr. 11, 80937 München, Germany

Materials

Equipment

LightCycler instrument (Roche Diagnostics, Mannheim, Germany)
LightCycler software vers. 3.3 (Roche Diagnostics)
LightCycler Carousel Centrifuge (Roche Diagnostics)
Oligo Primer analysis software (Molecular Biology Insights, Inc., Cascade, CO)

Reagents

Amplification Primers (TIB MOLBIOL, Berlin, Germany)
Hybridization Probes (TIB MOLBIOL)
Genomix kit (Fröbel, Lindau, Germany)

The following reagents were purchased from Roche Diagnostics:
LightCycler-DNA Master Hybridization Probes

Strains

Fifteen inactivated preparations of *Coxiella burnetii* strains (see Table 1) were kindly provided by G. Baljer (Giessen, Germany).

Procedure

Preparation of Template DNA

The concentration of a *Coxiella*-suspension of the reference strain Nine Mile had been determined microscopically. Total DNA was extracted using the Genomix kit and a series of dilutions was prepared. Aliquots containing 10^6, 10^5, 10^4, 10^3, 10^2, 10^1, 10^0, and 10^{-1} particles in a volume of 2 µl were analyzed by PCR.

In order to prepare DNA from cultured bacteria or from inactivated *C. burnetii* isolates, the suspensions were gently mixed with the same amount of lysis buffer

Table 1. Origin and host species of *Coxiella burnetii* isolates used in this study

Isolate	Origin	Host species
Nine Mile RSA493	USA	Tick
Priscilla Q177	USA	Goat
Scurry Q217	USA	Human
Dugway 5J108–111	USA	Rodent
Z2775	Germany	Cattle
Pohlheim	Germany	?
Max	Germany	?
Tiho 1	Germany	?
Namibia	Namibia	Goat
F-2	France	Human
F-4	France	Human
R1140	Russia	Human
CS-Florian	Slovak Republic	?
Gbud	Slovak Republic	Cattle
CS-R	Italy	Human

(50 mM Tris-HCL, 1 mM Na_2 EDTA, 0.5% Tween 20, pH 8.0) containing Proteinase K (50 µg/ml). After incubation at 56°C for 2 h the lysates were boiled for 15 min. We used 2 µl of the lysate directly for PCR.

Primer and Probe Design

A previously published sequence of the transposase gene was used for identifying *C. burnetii*-specific primer sequences [3]. This transposon-like element (GenBank Accession #M80806) exists in at least 19 copies in the *C. burnetii* Nine Mile genome, thereby being an ideal target for setting up a sensitive assay. As no further sequence entries with regard to this gene were deposited in GenBank or EMBL databases, we sequenced a 294-bp amplicon derived from 14 different *Coxiella* isolates (Table 2). An alignment was performed demonstrating a high degree of sequence homology (99%–100%) among *Coxiella* isolates. Candidate oligonucleotides serving as LightCycler hybridization probes for the sequence-specific detection of amplicons were chosen using the Oligo software to ensure comparable T_m values and GC contents within the range of 35%–60%, and to avoid self-complementarity and stable secondary structures.

LightCycler PCR

The following master mix was used for amplification and hybridization probe-based detection of the *Coxiella burnetii*-specific amplicons:

	Volume [µl]	[Final]
LightCycler-DNA Master Hybridization Probes	2	1×
$MgCl_2$ stock solution	2.4	4 mM
Primers (10 µM each)	1+1	0.5 µM
Hybridization probes (3 µM each)	1+1	0.15 µM
H_2O (PCR grade)	9.6	
Total volume	18	

Table 2. Oligonucleotides

Coxiella burnetii transposase gene (IS 1111a) (GenBank Accession #M80806)				
	Position	Length	GC (%)	T_m (°C)
Primers				
GTCTTAAGGTGGGCTGCGTG	219	20	60.0	66.86
CCCCGAATCTCATTGATCAGC	512 R	21	52.4	64.27
Product	219–512	294		
Probes				
GTTACTTTTGACATACGGTTTGACGTGCT-F	349 R	29	41.4	68.93
LCRed640- CGGACTGATCAACTGCGTTGGGAT	319 R	24	54.2	70.29

Precooled capillaries were loaded each with 18 µl master mix and 2 µl template DNA. The negative control sample was prepared by replacing the DNA template with PCR-grade water. The positive control sample was prepared by adding 2 µl of *C. burnetii* genomic DNA to the master mix. After a short centrifugation, the carousel with the capillaries was placed into the LightCycler rotor and amplification was started according to the following protocol:

- Initial denaturation for 2 min at 95°C
- Amplification

Parameter	Value		
Cycles	45		
Type	Quantification		
	Segment 1	Segment 2	Segment 3
Target temperature [°C]	95	55	72
Incubation time [s]	0	10	13
Temperature transition rate [°C/s]	20	20	20
Acquisition mode	None	Single	None
Gains	F1=1; F2=15; F3=30		

- Cooling for 2 min at 40°C

Results

Based on the sequence of a previously published *Coxiella* transposase gene, the selected set of primers and hybridization probes proved to be particularly valuable for the detection of *Coxiella* DNA. This system was examined to determine if it could simplify the diagnostic laboratory workflow. As with every modified or new PCR protocol, the sensitivity and specificity have to be examined before introducing it into routine testing.

Sensitivity PCR experiments were performed on serial dilutions of genomic DNA prepared from cultured *C. burnetii* organisms in order to determine the assay's lower limit of detection. A detection limit of one particle per PCR reaction was determined (Fig. 1).

Specificity The specificity of the PCR assay was evaluated with a collection of 15 cultured *C. burnetii* isolates from various geographic sources and hosts (Table 1). Fifteen Gram-negative or Gram-positive bacterial isolates other than *Coxiella* were also examined. All DNA preparations derived from the *Coxiella* isolates were amplified and detected, whereas none of the other bacterial strains examined was recognized. Experimental results from a panel of cultured bacteria are shown in Fig. 2. The curves generated by the LightCycler software allow a clear discrimination between the positive control (*C. burnetii* strain Nine Mile) and bacteria other than *Coxiella*.

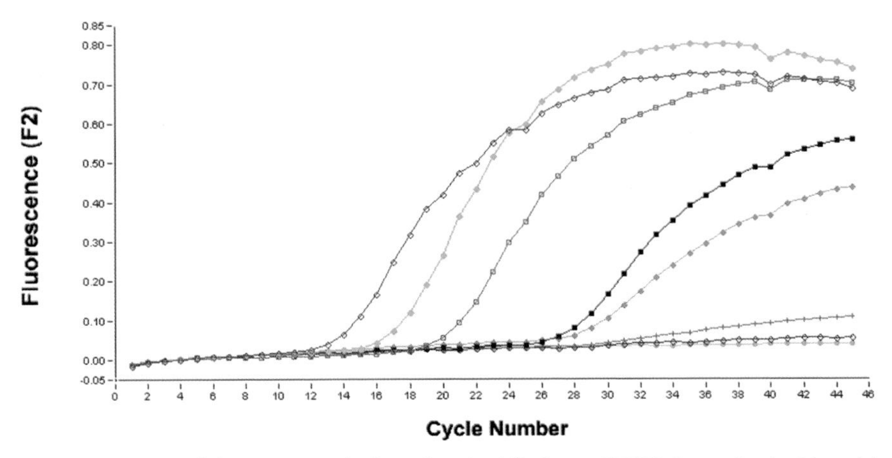

Fig. 1. Sensitivity of the PCR assay in detecting *Coxiella burnetii* DNA determined with serial dilutions of genomic DNA. For each dilution, the corresponding *Coxiellae* particle number present in the reaction mixture is given

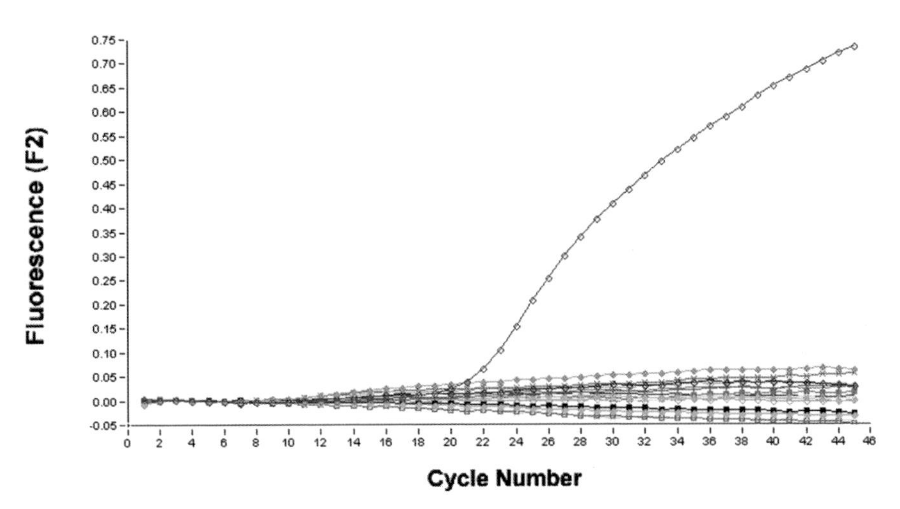

Fig. 2. Evaluation of the *Coxiella burnetii*-specific PCR assay with a representative set of bacterial strains other than *Coxiella. C. burnetii* strain Nine Mile was used as positive control

Comments

Previous methods for diagnostic PCRs had to include time-consuming and laborious procedures to prove the identity of the amplified product. The protocol described here simplifies the PCR workflow by a completely automated amplification and online detection procedure of sequence-specific probes. Unambiguous results are obtained in less than 30 min and as little as one *Coxiella* organism can be detected per assay. This was achieved by targeting a repetitive sequence that exists in up to 19 copies in the *C. burnetii* reference strain Nine Mile, thereby enhancing sensitivity. Applying this protocol could resolve a number of issues, which in classical detection methods are hampered by low sensitivity and an inherent danger for the laboratory investigator due to this highly infectious agent. This method might be applied for the detection of chronic Q fever, which in its most severe complication manifests as infective endocarditis [4]. In other cases of chronic Q fever, a number of different symptoms can be observed. The current debate involves whether the chronic form of Q fever occurs as a result of reactivation of latent infections. A second application would be to support epidemiological studies. It is well known that infections of ruminants with *C. burnetii* are widespread and often latent. The agent may be shed continuously or discontinuously via the milk or in large amounts in cases of abortion [5]. The risk of infections is increased if humans are exposed to these animals during parturition. *Coxiella* organisms may then persist in the local environment for weeks and months. However, little is known about how *C. burnetii* infections are maintained in nature and which animals or arthropods are involved.

References

1. Maurin M, Raoult D (1999) Q Fever. Clin Microbiol Rev 12:518–553
2. Fournier P et al. (1998) Diagnosis of Q fever. J Clin Microbiol 36:1823–1834
3. Willems H et al. (1994) Detection of *Coxiella burnetii* in cow's milk using the polymerase chain reaction. Zentralbl Veterinarmed [B] 41:580–587
4. Murray PR et al. (1999) In: ¦Manual of Clinical Microbiology, 7th edn. American Society for Microbiology, pp 815–820
5. Lorenz H et al. (1998) PCR Detection of *Coxiella burnetii* from different clinical specimens, especially bovine milk, on the basis of DNA preparation with a silica matrix. Appl Environ Microbiol 64:4234–4237

Molecular Identification of *Campylobacter coli* and *Campylobacter jejuni* Using Automated DNA Extraction and Real-Time PCR with Species-Specific TaqMan Probes

Lynn A. Cooper*, Luke T. Daum, Robert Roudabush, Kenton L. Lohman

Introduction

Campylobacter species are among the most common bacterial culprits in food and water associated incidents of acute gastroenteritis [1]. A particular problem in the livestock and dairy industries, *Campylobacter coli* and *Campylobacter jejuni* are the primary species within this group that pose a significant human health threat [2–4]. In the United States alone, the annual number of clinically confirmed cases of *C. coli* and *C. jejuni* number in the millions [5].

Because of their public health importance, we developed a real-time PCR assay which utilizes the Roche LightCycler instrument and species-specific TaqMan fluorescent probes to identify the presence of *Campylobacter coli* and *Campylobacter jejuni* DNA. In order to demonstrate their specificity for *C. coli* and *C. jejuni*, we tested the probes on a broad collection of other *Campylobacter* species and closely related organisms (i.e., nearest neighbors). For reproducible and reliable isolation of DNA from these organisms we have implemented a protocol using the Roche MagNA Pure. We then applied this protocol to the isolation of *C. coli* and *C. jejuni* DNA directly from stool samples. In this report, we present a brief synopsis of our experimental methods, data that illustrate the specificity of our probes, and experimental results that confirm the reliability of DNA extraction and amplification that was achieved by using the Roche MagNA Pure LC automated DNA extraction system coupled with the LightCycler PCR instrument.

Materials

Equipment

LightCycler instrument
Roche MagNA Pure LC System (Roche Molecular Biochemicals, Indianapolis, IN, USA)
Primer Express Software (Perkin Elmer Applied Biosystems, Foster City, CA, USA)

* Lynn A. Cooper (✉) (e-mail: lynn.cooper@brooks.af.mil)
Molecular Epidemiology Branch, AFIERA/SDE, Brooks AFB, 2601 West Gate Road, Suite 114, San Antonio, Texas 78235, USA

Reagents

Roche MagNA Pure DNA Isolation Kit I (Roche Molecular Biochemicals)
Roche LightCycler DNA Master Hybridization Probes Kit (Roche Molecular Biochemicals)

Procedure

Sample Preparation

A detailed description of our overall experimental design and techniques is to be reported elsewhere (Cooper et al., in preparation); however, a brief overview of our methodology is described here. Seed stocks of 14 *Campylobacter* species were obtained from American Type Culture Collection (ATCC, Rockville, MD, USA), and working stocks of each were generated in house using appropriate growth conditions specific to each organism. Additional *C. jejuni* isolates were obtained from culture confirmed clinical samples. The more distantly related bacterial strains used for cross-reactivity studies were obtained from the ATCC or from cultured material derived from clinical patient samples.

Test DNA was isolated from purified bacterial cultures, as well as directly from patient stool specimens. Automated DNA extractions were performed with the Roche MagNA Pure LC using the corresponding DNA Isolation Kit I in accordance with the manufacturer's instructions. The number of organisms present in any given sample was estimated by quantifying the amount of DNA extracted on a UV spectrophotometer. The number of genome copies present in the extracted sample was then calculated based on the amount of DNA detected and the organism's genome size.

Oligonucleotides

In brief, the assay uses a single set of PCR primers designated as CJCC-Forward and CJCC-Reverse to amplify the DNA of both *C. coli* and *C. jejuni*. At present, we cannot disclose the specific design of our oligonucleotide sequences; however, these primers were designed against conserved genetic regions common to both the *C. jejuni* protein binding gene and the ceuE gene of *C. coli*. Real-time fluorescent detection of target sequences is achieved by means of specific TaqMan probes. CJEJBP-TM1 is targeted specifically to detect only *C. jejuni* DNA and CCOLICEUE-TM1 is designed to specifically detect *C. coli* DNA target sequences. Each probe is labeled at the 5′ end with 6-carboxyfluorescein (FAM) and with 6-carboxytetramethylrhodamine (TAMRA) at the 3′end. The primers and probes were designed with the aid of Primer Express Software (Perkin Elmer Applied Biosystems, Foster City, CA, USA). Specifically, we used the TaqMan primer and probe design feature of this software package set to the default specifications. These included a probe annealing temperature that is approximately 10°C higher than the primer annealing temperature.

LightCyler PCR

All PCR reactions were carried out using reagents from the Roche LightCycler DNA Master Hybridization Probes kit and were performed by a LightCycler instrument. The following master mix recipe was used for all PCR amplifications.

Hybridization probe master mix for a 20-μl reaction volume:

	Volume [μl]	[Final]
HO	11.7	–
MgCl(25 mM)	3.2	5 mM
Hybridization mix ×10	2.0	1×
Forward primer (20 μM stock)	0.5	0.5 μM
Reverse primer (20 μM stock)	0.5	0.5 μM
TaqMan Probe (20 μM stock)	0.1	0.1 μM
Template DNA	2.0	Sample dependent

The amount of total DNA per sample extraction typically ranged from 2 to 35 ng/μl.

PCR amplification and probe detection of species-specific target sequences were accomplished using the following cycling conditions:

Parameter	Value	
Cycles	45	
Type	Quantification	
	Segment 1	Segment 2
Target temperature [°C]	95	60
Incubation time [s]	0	60
Temperature transition rate [°C/s]	20	20
Acquisition mode	None	Single
Gains	F1=1; F2=10; F3=10	

Results

Genetic material from representative isolates of various *Campylobacter* species was tested to demonstrate the specificity of our *C. coli* and *C. jejuni* species-specific TaqMan probes. A list of these *Campylobacter* species, as well as a partial listing of the nearest neighbor nucleic acids that were tested for cross-reactivity to each of the probes are shown in Table 1. In all cases, no amplification of non-target DNA was observed. These data strongly support that each of the two probes, one designed to detect only *C. coli* DNA and the other to detect only *C. jejuni* DNA, are each highly specific for their target organism. Furthermore, each probe can consistently amplify samples that contain less than 100 genomic copies of its respective target organism.

We also performed an experiment to assess the ability of the combined MagNA Pure automatic extraction and LightCycler PCR methods to consistently yield reproducible results. The DNA from four sets of identically aliquoted and diluted samples (n=16) of a cultured *C. jejuni* isolate were independently extracted and amplified. The results of this experiment are shown in Fig. 1. Over a four log dynamic range on replicate samples containing 1.2×10, 10, 10, or 10 colony form-

Table 1. *Campylobacter* species and nearest neighbor nucleic acids tested for cross-reactivity with *Campylobacter coli* and *Campylobacter jejuni* specific TaqMan probes. In all cases, no cross-reactivity was observed with these control organisms

Cross-Reactivity-Reference-Panel	
Campylobacter coli	*Campylobacter concicus*
Campylobacter curvus	*Campylobacter fetus*
Campylobacter gracilis	*Campylobacter helveticu*
Campylobacter hyoilei	*Campylobacter jejuni*
Campylobacter hypointestinalis	*Campylobacter lari*
Campylobacter mucosalis	*Campylobacter rectus*
Campylobacter sputorum	*Campylobacter upsaliensis*
Citrobacter	*Eschericia ssp.*
Enterobacter	*Proteus*
Serratia	*Salmonella ssp.*
Shigella ssp.	Human total genomic DNA

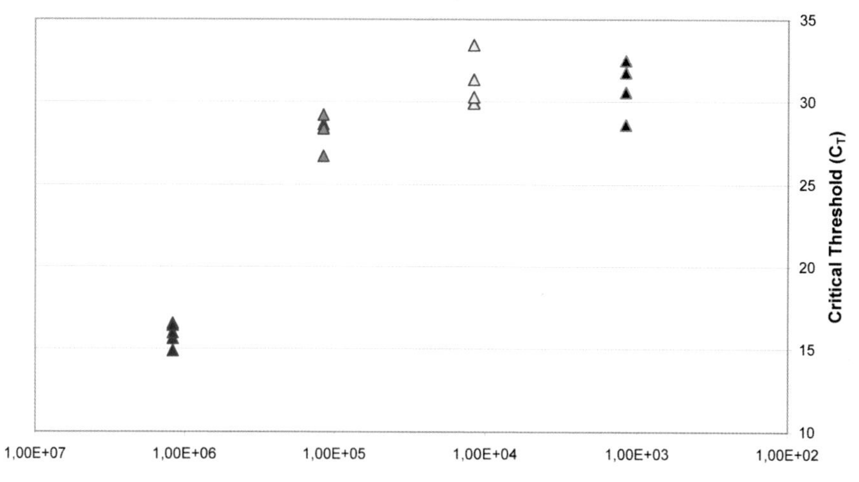

Fig. 1. Reproducibility of results obtained by MagNA Pure extraction and LightCycler amplification of *C. jejun* DNA. Each *triangle* represents one of 16 independently extracted and amplified *C. jejuni* samples

ing units, we observed a tight clustering of threshold cycle values within each of these four groups. The threshold cycle (C_T) is defined as the PCR cycle where a significant increase in fluorescence was first observed [6]. This value is normalized against background fluorescence. These data clearly illustrate that automated

DNA extraction methods paired with subsequent LightCycler amplification techniques can yield highly consistent results.

Comments

In this brief report, we describe a rapid and reproducible method for the isolation and analysis of DNA from cultured *Campylobacter coli* and *Campylobacter jejuni*. The technique which specifically identifies these pathogens using TaqMan real-time PCR fluorescent probes on the Roche LightCycler instrument has wide-ranging applications. Furthermore, the assay is highly specific, and no probe cross-reactivity was observed when the reagents were tested against an extensive panel of closely related *Campylobacter* strains and other microbial agents. The combined speed, sensitivity, and specificity of this assay system make it a powerful diagnostic tool.

Rapid nucleic acid based diagnostic procedures for infectious organisms, such as those we describe here for *C. coli* and *C. jejuni*, provide unmatched sensitivity, specificity, and speed [7]. However, they often suffer from slow manual sample preparation methods which greatly reduces the efficacy of high throughput applications. The combination of the MagNA Pure and LightCycler provides virtually walk-away sample preparation and reproducible high throughput analyses of pathogenic organisms. The future applications of these technologies to large scale screening efforts in clinical laboratories abound. By demonstrating the feasibility of these techniques for the detection of *C. coli* and *C. jejuni* DNA, we have put forth a model on which to base the future development of similarly efficient assays for numerous etiological agents of acute human illness.

References

1. Allos BM (2001). *Campylobacter jejuni* infections: update on emerging issues and trends. Clin Infect Dis 32:1201–1206
2. Lucey B, Crowley D, Moloney P, Cryan B, Daly M, O'Halloran F, Threlfall EJ, Fanning S (2000). *Campylobacter* ssp. of human and animal origin. Emerg Infect Dis 6(1): 50–55
3. Wesley IV, Wells SJ, Harmon KM, Green A, Schroeder-Tucker L, Glover M, Siddique I (2000). Fecal shedding of *Campylobacter* and *Arcobacter* ssp. in dairy cattle. Appl Environ Microbiol 66:1994–2000
4. Blaser MJ (1997). Epidemiologic and clinical features of *Campylobacter jejuni* infections. J Infect Dis 176 [Suppl 2]: S103–105
5. Altekruse SF, Stern NJ, Fields PI, Swerdlow DL (1999). *Campylobacter jejuni* – an emerging foodborne pathogen. Emerg Infect Dis 5:28–35
6. LightCycler Owner's Manual (1998). Roche Molecular. Version 1.2
7. Lai-King NG, Kingombe CIB, Yan W, Taylor DE, Hiratsuka K, Malik N, Garcia MM (1997). Specific detection and confirmation of *Campylabacter jejuni* by DNA hybridization and PCR. Appl Environ Microbiol 63:4558–4563

Rapid and Specific Detection of *Plesiomonas shigelloides* Directly from Stool by LightCycler PCR

Jin Phang Loh, Eric Peng Huat Yap*

Introduction

Plesiomonas shigelloides is a Gram-negative bacterium which has been implicated as an agent of human gastroenteritis [1] and is recognized as being an important cause of seafood-associated outbreaks [2, 3]. A study in Japan ranked *P. shigelloides* first (66.7%) in a group of enteropathogens causing travellers' diarrhoea. Septicaemia, cellulitis, meningitis, cholecystitis and endophthalmitis due to *P. shigelloides* have also been documented among immunocompromised patients. The mortality rate of these patients associated with *Plesiomonas* septicemia is high.

In healthy individuals, *Plesiomonas shigelloides* gastroenteritis is usually a mild self-limiting disease with fever, chills, abdominal pain, nausea, diarrhoea or vomiting. The infectious dose is presumed to be quite high, at least greater than one million organisms. Bacteriological methods for the isolation and identification of *Plesiomonas* are long and laborious [4]. Since most laboratories concentrate on recovery of *Salmonella*, *Shigella*, *Escherichia coli* and other classical enteric pathogens, *P. shigelloides* may be overlooked during routine culture of stool samples. Furthermore, there is no selective media for the concurrent culture of *P. shigelloides* and ability to differentiate it from the *Aeromonas* species and other natural microflora [1, 4]. Correct diagnosis therefore requires stringent methods of laboratory culture and identification. Stool culture, however, still serves as the definitive means for identifying an enteric infection due to *P. shigelloides*. Culture on semi-selective media such as SS agar or XLD agar followed by identification by biochemical tests usually requires 48–72 h.

Polymerase chain reaction (PCR) techniques have become increasingly important as a means for detection of different bacterial pathogens. However, these techniques have not been routinely applied in clinical microbiology as many DNA methods largely depend on traditional culture procedures for enrichment of bacteria and hence are not truly "rapid methods" [5]. Use of PCR directly on stool specimens is even more limited because of the presence of unidentified PCR inhibitors in such specimens. Several methods have been described for removal or inactivation of PCR inhibitors in stool specimens, most of which are laborious or inefficient [6].

* Eric Peng Huat Yap (✉) (e-mail: nmiv3@nus.edu.sg)
Defence Medical Research Institute, Defence Science Technology Agency,
MD11 Clinical Research Centre, #02-04, 10 Medical Drive, Singapore 117597

González-Rey et al. [7] recently developed a PCR assay for the specific detection of *Plesiomonas shigelloides* based on the 23S rRNA gene. The assay was shown to be specific and detected isolates from aquatic environments, animals and human diarrhoea cases. In this study, we developed another method which combined the use of a commercial kit for rapid DNA preparation from stool (QIAamp DNA Stool Mini Kit) with a real-time capillary FRET PCR assay developed for specific detection of *Plesiomonas shigelloides*.

The real-time FRET PCR assay developed in this study utilizes SYBR Green I together with LCRed640-labelled probe for detection of PCR products. SYBR Green I binds to double-stranded DNA and is a generic indicator of amplification. SYBR Green I has an excitation maximum near that of fluorescein, thus enabling it to be a generic donor in FRET to excite LCRed640, which is labelled on a DNA probe designed specifically to bind to the correct PCR product. A similar approach was demonstrated previously by Cane et al. [8] where SYBR Green I was used to excite a Cy5-labelled probe in a real-time PCR assay. This results in a rapid and specific assay for *P. shigelloides* that can be used for detection directly from stool without the need for traditional culture.

Materials

Equipment

LightCycler instrument (Roche Diagnostics, Mannheim, Germany)
LightCycler software vers. 3.39 (Roche Diagnostics)
LightCycler Capillaries (Roche Diagnostics)
LightCycler Centrifuge Adapters (Roche Diagnostics)
LightCycler Cooling Block (Roche Diagnostics)
Oligo 5.0 software (Molecular Biology Insights Inc., Cascade, CO)
Primer Select software (DNAStar Inc., Madison, WI)

Reagents

Amplification Primers (GENSET Singapore Pte Ltd., Singapore)
Hybridization probe (GENSET Singapore Pte Ltd)
QIAamp DNA Mini Kit (Qiagen, Hilden, Germany)
QIAamp DNA Stool Mini Kit (Qiagen)
Platinum *Taq* DNA polymerase (Invitrogen Corp., Carlsbad, CA)
Bovine Serum Albumin (BSA; Roche Diagnostics)
SYBR Green I (Molecular Probes, Eugene, OR)

Procedure

Sample preparation

Genomic DNA from various bacterial species used for specificity testing was extracted using the QIAamp DNA Mini Kit. DNA from stool samples were extracted using the QIAamp DNA Stool Mini Kit.

Extraction of DNA from Stool Using Qiagen DNA Stool Mini Kit

Certain modifications to the manufacturer's protocol for isolation of DNA from stool for pathogen detection were made. Briefly, 400 µl of stool sample was mixed with 1 ml of ASL buffer and thoroughly homogenized. The stool suspension was then heated to 95C for 5 min before centrifugation at 14,000 rpm for 1 min. We then pipetted 1.2 ml of the supernatant into a 2-ml microcentrifuge tube and added one InhibitEX tablet, followed by vigorous vortexing to mix before leaving at room temperature for 1 min. The sample was then centrifuged at 14,000 rpm for 3 min, after which 200 µl of supernatant was transferred to a fresh microcentrifuge tube. We added 15 µl of Proteinase K to the supernatant, which was transferred and mixed, followed by the addition of 200 µl AL buffer and vortexing. The sample was then incubated at 70C for 10 min after which 200 µl of ethanol was added and mixed. The whole mixture was then loaded onto a QIAamp spin column and centrifuged at 14,000 rpm for 1 min. The QIAamp spin column was placed in a new 2-ml collection tube and 500 µl AW1 buffer added to the spin column. The spin column was then centrifuged at 14,000 rpm for 1 min and transferred to another new 2-ml collection tube. We added 500 µl of AW2 buffer and centrifuged it at 14,000 rpm for 3 min. The QIAamp spin column was placed in a new 1.5-ml microcentrifuge tube and 100 µl AE buffer was pipetted directly onto the QIAamp membrane. This was incubated for 1 min at room temperature and centrifuged at 14,000 rpm for 1 min to elute the DNA. Then, 1.5 µl of the eluate was used in the PCR assay.

Oligonucleotides

González-Rey et al. [7] developed a PCR assay for the specific detection of *P. shigelloides* using 23S rDNA as a target sequence. We also designed primers complementary to 23S rDNA resulting in a shorter amplicon of 112 bp for rapid thermocycling using the LightCycler system. Primers were designed using Oligo 5.0 software. A LCRed640-labelled probe was designed using Primer Select Software (Table 1).

Table 1. Oligonucleotides

Plesiomonas shigelloides 23S RNA Gene (GenBank Accession #X65487)					
	Position	Length	GC (%)	T_m	Purity
Primers:					
AGCGCCTCGGACGAACACCTA	827	21	61.9	71.6	1.0
GTGTCTCCCGGATAGCAC	939R	18	61.1	62.9	1.0
Product:	827–939	112			
Hybridization probe:					
LCRed640-GGTAGAGCACTGT-TAAGGCTAGGGGGTCATC-P	852	31	54.8	74.0	1.25

LightCycler PCR

The following master mix was used for amplification and detection of *P. shigelloides* amplicons:

	Volume [µl]	[Final]
Platinum Taq buffer (10×)	1	1×
dNTPs (2 mM each)	1	0.2 mM each
Primers (10 µM each)	0.3+0.3	0.3 µM each
BSA (5 µg/µl)	1	0.5 µg/µl
$MgCl_2$ (25 mM)	1.6	4 mM
Probe (20 µM)	1	0.2 µM
SYBR Green I (10×)	1	1×
Platinum Taq (5 U/µl)	0.1	0.5 U
H_2O	1.2	

A 1.5-µl volume of sample DNA was added to 8.5 µl of the LightCycler master mix in the LightCycler capillaries. The capillaries were closed, centrifuged in a microcentrifuge using the LightCycler adapters and placed in the LightCycler sample carousel.

The following protocol was used to detect and identify the *P. shigelloides* species by amplifying a segment of the 23S RNA gene:

- Denaturation for 1 min at 95°C
- Amplification

Parameter	Value		
Cycles	70		
Type	Quantification		
	Segment 1	Segment 2	Segment 3
Target temperature [°C]	90	70	72
Incubation time [s]	0	4	5
Temperature transition rate [°C/s]	20	20	20
Acquisition mode	None	Single	None

A denaturation temperature of 90C was used instead of 95C because short PCR product yields are significantly improved by lower denaturation temperatures [9]. A 3-step thermocycler profile is applied instead of a 2-step profile to maintain the high specificity required.

- Melting Curve Analysis

Parameter	Value		
Cycles	1		
Type	Melting curve analysis		
	Segment 1	Segment 2	Segment 3
Target temperature [°C]	90	70	94
Incubation time [s]	0	20	0
Temperature transition rate [°C/s]	20	20	0.2
Acquisition mode	None	None	Continuous

- Cooling for 2 min at 40°C

A sensitivity test was done by carrying out PCR on serially diluted *P. shigelloides* DNA. Specificity of the PCR assay developed was determined by assaying against 27 other bacterial species of which several are common causes of gastroenteritis.

Gel Electrophoresis of PCR Products

The LightCycler PCR products were separated by gel electrophoresis to verify the presence of specific products. The amplified 23S RNA PCR products were removed from the capillary by reverse centrifugation into a 0.6-ml Eppendorf tube. A 2-μl volume of gel loading buffer was added and the sample was electrophoretically analyzed on a 2% agarose gel.

Results

Sensitivity

The analytical sensitivity of the assay was determined to be approximately 100 fg (about 20 organism equivalents; see Fig. 1, 2) with weak amplification observed for 10 fg by using serial dilutions of *Plesiomonas shigelloides* genomic DNA isolated from cultured *P. shigelloides* organisms.

Specificity

The specificity of the assay was evaluated against 27 other bacterial species (Table 2). The results show amplification of only the *Plesiomonas shigelloides* reference strains (Figs. 1, 2). None of the closely related bacterial species such as *Aeromonas hydrophila, A. sobria* and *Vibrio cholerae* gave a positive amplification signal in LightCycler and agarose gel analysis. Although the sequence of the hybridization probe is complementary to a number of *Vibrio* spp., none of the *Vibrio* spp. tested gave a postive signal. This indicates the specificity of the selected primer sequences.

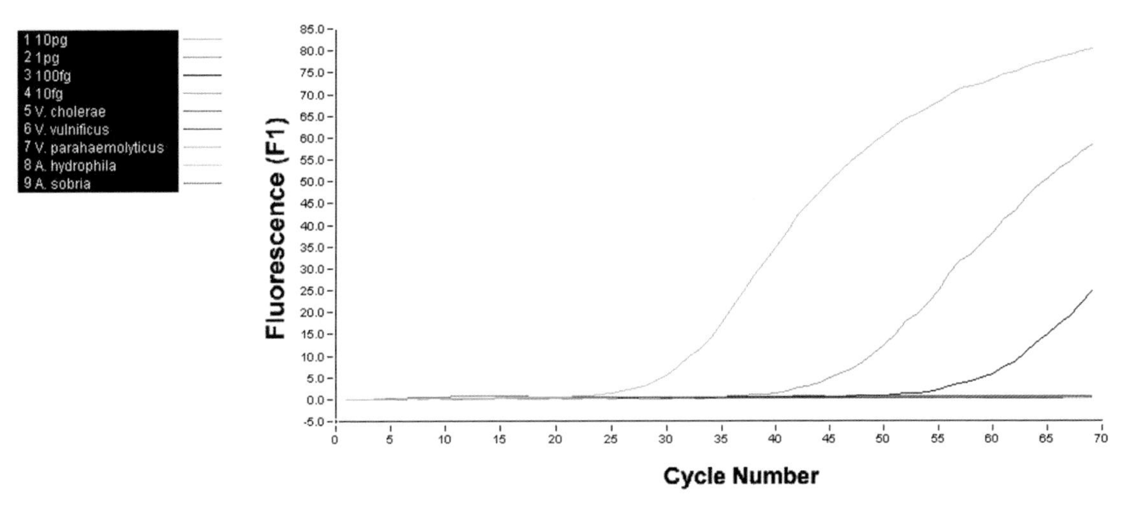

Fig. 1. Analytical Sensitivity and specificity of *Plesiomonas shigelloides* FRET PCR using SYBR Green I fluorescence detection on the F1 channel of the LightCycler

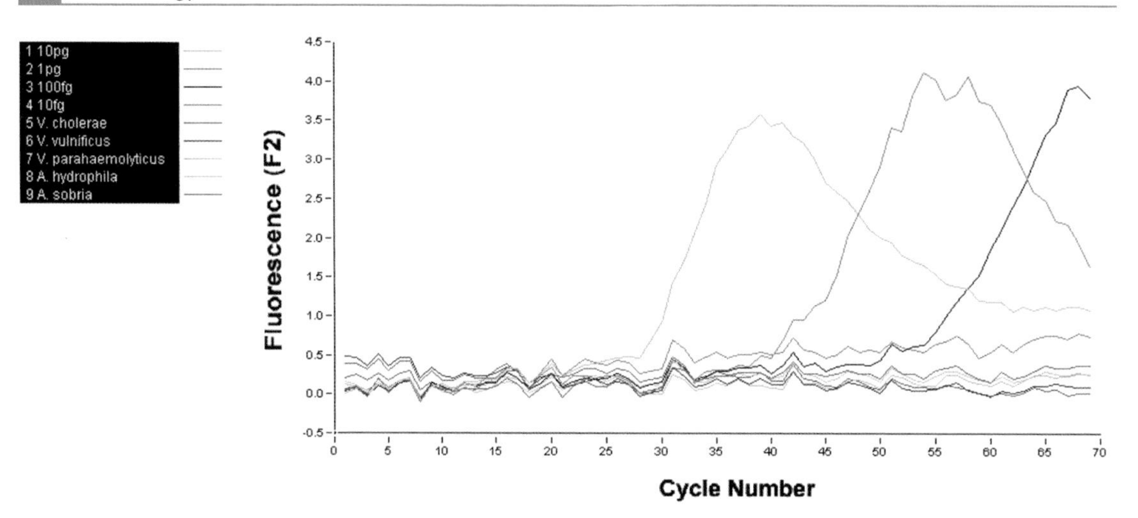

Fig. 2. Analytical Sensitivity and specificity of *Plesiomonas shigelloides* FRET PCR using LC640 fluorescence detection on the F2 channel of the LightCycler

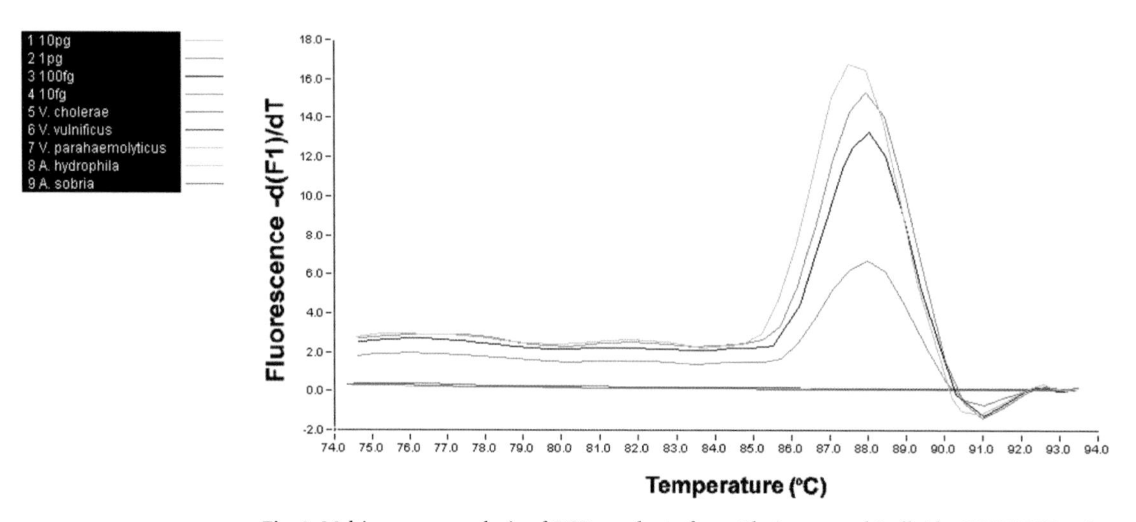

Fig. 3. Melting curve analysis of PCR products from *Plesiomonas shigelloides* FRET PCR using SYBR Green I fluorescence detection on the F1 channel of the LightCycler

Melting Temperature

The presence or absence of specific PCR products was determined by the melting curve analysis in channel F1. The mean T_m values for the reference strains of *Plesiomonas shigelloides* were 88°C (Fig. 3), easy to differentiate from primer dimers which has a T_m of 84°C (not shown). This is possible because the LightCycler is capable of differentiating two PCR products whose T_ms differ by less than 2°C

Table 2. Specificity of *Plesiomonas shigelloides* PCR assay

Organism	Strain number	Source[a]	PCR result
P. shigelloides	W2/3/94	SGH	+
P. shigelloides	M1C 3/94	SGH	+
Aeromonas hydrophila	W3/4/98	SGH	–
Aeromonas sobria	MIB 8/96	SGH	–
Bacillus subtilis	NCTC 10073	DMRI	–
Bacillus cereus	NCTC 2599	DMRI	–
Campylobacter jejuni	ATCC 29428	DMRI	–
Clostridium botulinum	NCTC 8266	DMRI	–
Clostridium perfringens	ATCC 13124	DMRI	–
Enterococcus faecalis	ATCC 19433	DMRI	–
Enterococcus faecium	ATCC 19434	DMRI	–
Escherichia coli	ATCC 25922	DMRI	–
Escherichia coli O157:H7	NCTC 12079	DMRI	–
Helicobacter pylori	DM 8505/00	SGH	–
Listeria monocytogenes	NCTC 10357	DMRI	–
Proteus mirabilis	PM1	DMRI	–
Pseudomonas aeruginosa	ATCC 27853	DMRI	–
Salmonella paratyphi A	W2/1/92	SGH	–
Salmonella paratyphi B	W2/1/91	SGH	–
Salmonella paratyphi C	W2/3/89	SGH	–
Salmonella typhi	NCTC 8393	DMRI	–
Shigella dysenteriae	NCTC 9347	DMRI	–
Shigella flexneri	ATCC 12072	DMRI	–
Shigella sonnei	ATCC 25931	SGH	–
Staphylococcus aureus	NCTC 25923	DMRI	–
Vibrio cholerae	AS 4/96	SGH	–
Vibrio parahaemolyticus	NCTC 10903	DMRI	–
Vibrio vulnificus	NCTC 11066	DMRI	–
Yersinia enterocolitica	ATCC 9610	DMRI	–
Human DNA	SA1	DMRI	–

–, negative; +, positive.

[a] SGH, Singapore General Hospital, Singapore; DMRI, Defence Medical Research Institute, Singapore.

[10]. Melting curve analysis in channel F2 was not useful as SYBR Green I dissociation masks the specific LC Red 640 signal decrease.

Applications

The PCR assay combined with the use of the QIAamp DNA Stool Mini Kit for rapid detection and presumptive diagnosis was carried out successfully on a collection of 20 stool samples collected from patients suffering from diarrhoea. One of the samples tested positive for *P. shigelloides* by LightCycler PCR. An attempt was then made to isolate the organism from stool by plating on modified SS agar and XLD agar after enrichment. The stool culture was successful and the isolate verified using the biochemical assay API 20E.

Comments

Three levels of verification are possible in this FRET PCR assay. Firstly, an increase in SYBR Green I fluorescence detected in the F1 channel indicates the presence of specific PCR products being formed from the two specific primers (Fig. 1). In our specificity test against the other 27 bacterial species, no amplification products were observed (results not shown) for any of the other bacterial species. Secondly, increased fluorescence observed from LCRed640 detection channel (F2) indicates the labelled probe is binding to the correct PCR product and being excited by SYBR Green I, which binds to the double-stranded PCR product (Fig. 2). The characteristic "hook" effect for FRET PCR was also observed at the end of the exponential phase (Fig. 2). This is caused by competition between (a) the reassociation of the single strands of PCR product after denaturation and (b) the binding of the hybridization probes to target in the late PCR phase. The PCR process amplifies both template strands, so while the amounts of PCR product are high, the two strands reassociate faster than the hybridization probes can bind to their target [11]. Thirdly, the melting curve at the end of the PCR gives an indication of the T_m of the PCR product (Fig. 3). Primer dimers, if any, are easily differentiated from actual PCR product by their different T_ms.

Applications

A combination of the LightCycler PCR assay with the QIAamp DNA Stool Mini Kit provides rapid detection and presumptive diagnosis to be carried out within 3 h. Direct detection of *Salmonella* species and *Campylobacter jejuni* from stool of diarrhoea patients have also been demonstrated in our laboratory (results not shown). Advantages of this system include long-term storage of purified DNA from stool samples for future testing if required. With traditional culture techniques, survival of fastidious bacteria such as *Campylobacter jejuni* or unusual bacteria such as *Plesiomonas shigelloides* in stool samples for long periods of time may be limited and a second sampling at 24 or 48 h may or may not yield positive cultures. This is especially true if initial screens for common enteropathogens are unsuccessful.

The availability of a rapid PCR assay also assists in the screening process especially in gastroenteritis outbreak scenarios where time is crucial. For example, a positive PCR test for *Plesiomonas shigelloides* would make one examine the SS agar and XLD agar plates used for stool culture more closely or attempt stool culture on special selective or differential agar plates and special conditions. In summary, we have described in this paper, a sensitive, specific and rapid assay requiring only 3 h or less for direct detection of *Plesiomonas shigelloides* from stool.

References

1. Khardori N, Fainstein V (1988) *Aeromonas* and *Plesiomonas* as etiological agents. Ann Rev Microbiol 42:395–419
2. Rutala WA, Sarubbi FA, Finch CS, MacCormack JN, Steinkraus GE (1982) Oyster-associated outbreak of diarrhoeal disease possibly caused by *Plesiomonas shigelloides* (Letter). Lancet 1:739
3. Tsukamoto T, Kinoshita Y, Shimada T, Sakazaki R (1978) Two epidemics of diarrhoeal disease possibly caused by *Plesiomonas shigelloides*. J Hy. 80:275–280
4. Huq A, Aktar A, Chowdhury MAR, Sack DA (1991) Optimal growth temperature for isolation of *Plesiomonas shigelloides* using various selective and different agars. Can J Microbiol 37:800–802
5. Olsen JE, Aabo S, Hill W, Notermans S, Wernars K, Granum PE, Popovic T, Rasmussen HN, Olsvik Ø (1995) Probes and polymerase chain reaction for detection of food-borne bacterial pathogens. Int J Food Microbiol 28:1–78
6. Lou QY, Chong SKF, Fitzgerald JF, Siders JA, Allen SD, Lee CH (1997) Rapid and effective method for preparation of fecal specimens for PCR assays. J Clin Microbiol 35:281–283
7. González-Rey C, Svenson SB, Bravo L, Rosinsky J, Ciznar I, Krovacek K (2000) Specific detection of *Plesiomonas shigelloides* isolated from aquatic environments, animals and human diarrhoeal cases by PCR based on 23S rRNA gene. FEMS Immun Med Microbiol 29:107–113
8. Cane PA, Cook P, Ratcliffe D, Mutimer D, Pillay D (1999) Use of real-time PCR and fluorimetry to detect lamivudine resistance-associated mutations in hepatitis B virus. Antimicrob Agents Chemo 43:1600–1608
9. Yap EPH, McGee JOD (1991) Short PCR product yields improved by lower denaturation temperatures. Nucl Acids Res 19:171–1714
10. Ririe KM, Rasmussen RP, Wittwer CT (1997) Product differentiation by analysis of DNA melting curves during the polymerase chain reaction. Anal Biochem 245:154–160
11. Roche Molecular Biochemicals (1999) Decrease of fluorescent signal in the plateau phase of a LightCycler PCR hook effect. Roche Molecular Biochemicals Technical Note No. LC8/99

Quantification of Thermostable Direct Hemolysin-Producing *Vibrio parahaemolyticus* from Foods Assumed to Cause Food Poisoning Using the LightCycler Instrument

Yoshito Iwade*, Akinori Yamauchi, Akira Sugiyama, Osamu Nakayama, Norichika H. Kumazawa**

Introduction

Vibrio parahaemolyticus isolated from the feces of patients with food poisoning contains in most cases either thermostable direct hemolysin (TDH) and/or TDH-related hemolysin (TRH), both important pathogenic factors carried by this microorganism [1–3]. However, most organisms from brackish water, estuary mud or fish that cause food poisoning have neither the TDH production gene (*tdh*) nor the TRH production gene (*trh*) [1]. This seems to be because TDH-producing *V. parahaemolyticus* occurs in only small amounts in nature or foods compared to other *Vibrio* spp. In addition, it is difficult to grow in artificial media, since TDH-producing *V. parahaemolyticus* is in the viable but nonculturable (VNC) state due to nutrient limitation, low temperature, etc. [4, 5]. Therefore, TDH-producing *V. parahaemolyticus* is rarely isolated from foods assumed to cause food poisoning that have a high probability of containing TDH-producing *V. parahaemolyticus*. Accordingly, it is very difficult to quantitate TDH-producing *V. parahaemolyticus* by culture. In order to estimate the content of TDH-producing *V. parahaemolyticus* in suspect foods, we have developed a LightCycler assay to quantitate *tdh*.

Materials

Equipment

LightCycler instrument (Roche Diagnostics, Mannheim, Germany)
LightCycler software vers. 3 (Roche Diagnostics)
LightCycler Capillaries (Roche Diagnostics)
LightCycler Centrifuge Adapters(Roche Diagnostics)
Centrifuge (Tomy Seiko Co.,Ltd, Tokyo, Japan)

Reagents

Amplification Primers (Nihon Gene Research Lab's Inc., Sendai, Japan)
Hybridization Probes (Nihon Gene Research Lab's Inc., Sendai, Japan)
Brain Heart Infusion broth (Eiken Chemical Co.,Ltd, Tokyo, Japan)

* Yoshito Iwade, (✉) (e-mail: iwadey00@pref.mie.jp)
Mie Prefectural Institute of Public Health and Environmental Sciences, 3690-1, Sakura-cho, Yokkaichi-city, Mie-prefecture, 512-1211, Japan
** Norichika H. Kumazawa, Tropical Biosphere Research Center, University of the Ryukyus, 1, Senbarn, Nishihara-Cho, Nakagami-Gun, Okinawa-prefecture, 903-0213, Japan

Salt Polymyxin broth (Nissui Pharmaceutical Co.,Ltd, Tokyo, Japan)
ISOPLANT (Nippon Gene, Tokyo, Japan)

The following reagents were purchased from Roche Diagnostics:
LightCycler-DNA Master SYBR Green I
LightCycler-DNA Master Hybridization Probes

Procedure

Optimum Magnesium Chloride Concentration

The optimum $MgCl_2$ concentration was determined by using SYBR Green I. As shown in the following table, reactions with $MgCl_2$ concentrations from 2 mM to 5 mM were prepared at intervals of 1 mM.

SYBR Green I Assay for $MgCl_2$ optimization (volumes given in ul).

	2 mM	3 mM	4 mM	5 mM
LightCycler-DNA Master SYBR Green I	2.0	2.0	2.0	2.0
$MgCl_2$ (25 mM)	0.8	1.6	2.4	3.2
Primers (100 μM each)	0.2+0.2	0.2+0.2	0.2+0.2	0.2+0.2
H_2O (PCR grade)	14.8	14.0	13.2	12.4
Template DNA	2.0	2.0	2.0	2.0
Final volume [μl]	20.0	20.0	20.0	20.0

DNA extracted from a TDH-producing *V. parahaemolyticus* O3:K6 No.99273 strain (7 log CFU/mL) using ISOPLANT was added in 2-μl portions and amplification was carried out at the following thermocycle conditions:

- Denaturation for 1 min at 95°C
- Amplification

Parameter	Value		
Cycles	40		
Type	Quantification		
	Segment 1	Segment 2	Segment 3
Target temperature [°C]	95	55	72
Incubation time [s]	0	5	10
Temperature transition rate [°C/s]	20	20	20
Acquisition mode	None	None	Single

As Table 1 shows, we used a forward primer and a reverse primer designed by Nishibuchi et al. [6].

Formation of Standard Curve

The TDH-producing *V. parahaemolyticus* O3:K6 No.99273 strain isolated from the feces of a patient with food poisoning was cultured in 3% NaCl containing Brain Heart Infusion broth (BHIB) at 37°C for 18 h. The concentration of viable

Table 1. Oligonucleotides

(GenBank Accession X54341)				
	Position	Length	GC (%)	T_m (°C)
Primers				
GGTACTAAATGGCTGACATC	351	20	45.0	56.0
CCACTACCACTCTCATATGC	602	20	50.0	57.1
Product	351–602	251		
Probes				
ATGACCGTGCTTATAGCCAGAC-F	440	22	50.0	63.5
LCRed640-CCGCTCCATTGTATAGTCTTT	396	22	44.5	62.1

organisms was estimated mixing 10-fold serial dilutions with 3% NaCl containing Brain Heart Infusion agar, culturing at 37°C for 24 h, and counting colonies. DNA was also extracted from each dilution of the BHIB culture using ISOPLANT and dissolved in 50 µl of TE buffer. A quantity of 2 µl of the DNA extracted from each dilution was added to 18 µl of the following master mix and amplified under the conditions specified above:

	Volume [µl]	[Final]
LightCycler-DNA Master Hybridization Probes	2.0	1 x
MgCl (25 mM)	2.4	4 mM
Primers (100 µM each)	0.2 + 0.2	1 µM
Hybridization probes (50 µM each)	0.1 + 0.1	0.25 µM
H_20 (PCR grade)	13.0	
Total volume	18	

Treatment of Samples

Two samples causing food poisoning were studied, an omelet and a raw squid. TDH-producing *V. parahaemolyticus* O3:K6 had been previously isolated from both samples. Ten percent emulsions of the omelet and squid were prepared in 3% NaCl. From 10 ml of the omelet emulsion, DNA was directly extracted with ISOPLANT without culture. Both emulsions were also cultured by adding nine parts Salt Polymyxin broth and incubating at 37°C for 18 h. From 1-ml portions of these samples, DNA was extracted using ISOPLANT and dissolved in 50 µl of TE buffer. To complete the amplification mixtures, 18 µl of master mix and 2 µl of the corresponding template DNA were added to each capillary.

Results

Optimum Magnesium Chloride Concentration

Figure 1 shows the amplification curves at various $MgCl_2$ concentrations. The amplification efficiency is elevated with an increase in the $MgCl_2$ concentration. Since no large difference was observed between 4 mM and 5 mM, a $MgCl_2$ concentration of 4 mM was used throughout this study.

Quantification of TDH-Producing V. Parahaemolyticus in Foods

The TDH-producing *V. parahaemolyticus* O3:K6 No. 99273 strain was cultured in 3% NaCl containing Brain Heart Infusion Broth for 18 h and then serially diluted in ten-fold steps. Figure 2 shows the standard curve obtained. The standard curve shows good linearity (i.e., slope: –3.5, intercept: 44.5, error: 0.092).

As shown in Fig. 3, the DNA extracted from 10 ml of the 10% emulsion of the omelet with a 3% NaCl solution generated specific amplicons detectable from the 27th cycle on. The DNA extracted from 1 ml of the sample obtained by culturing the same omelet in nine times as much Salt Polymixin broth at 37°C for 18 h generated specific amplicons detectable from the 17th cycle on. The DNA extracted from the sample obtained by culturing the raw squid generated specific amplicons detectable from the 16th cycle on. The content of TDH-producing *V. parahaemolyticus* was estimated from the crossing point of each sample and the cali-

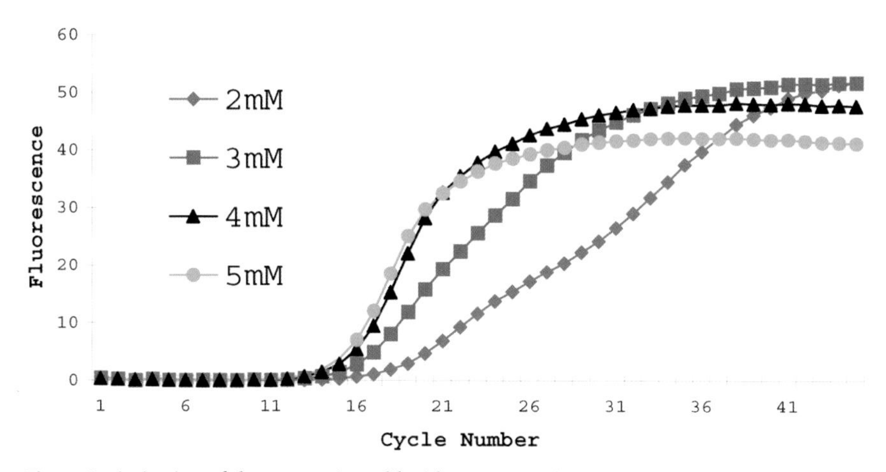

Fig. 1. Optimization of the magnesium chloride concentration

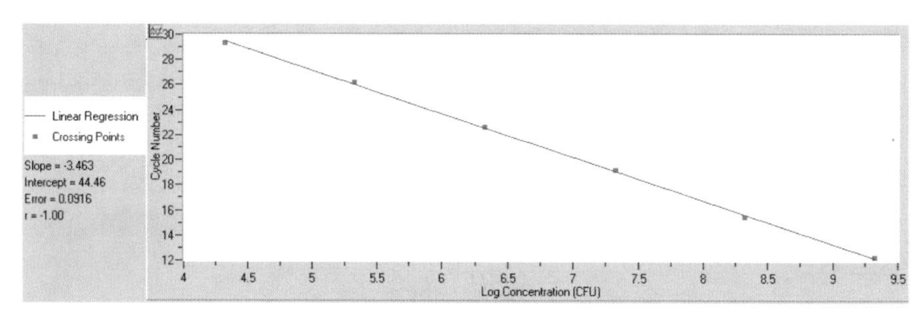

Fig. 2. Standard curve for the thermostable direct hemolysin gene

bration curve. The omelet sample, the omelet culture, and the raw squid culture, resulted in 3.4 logCFU/g (crossing point: 29.1 cycles), 7.7 logCFU/ml (crossing point: 17.7 cycles) and 7.1 logCFU/ml (crossing point: 19.8 cycles), respectively. Figure 4 shows the melting curves of the TDH-producing *V. parahaemolyticus* used in forming the calibration curve and the samples. A bimodal curve with peaks at 66°C and 59°C was obtained in each case.

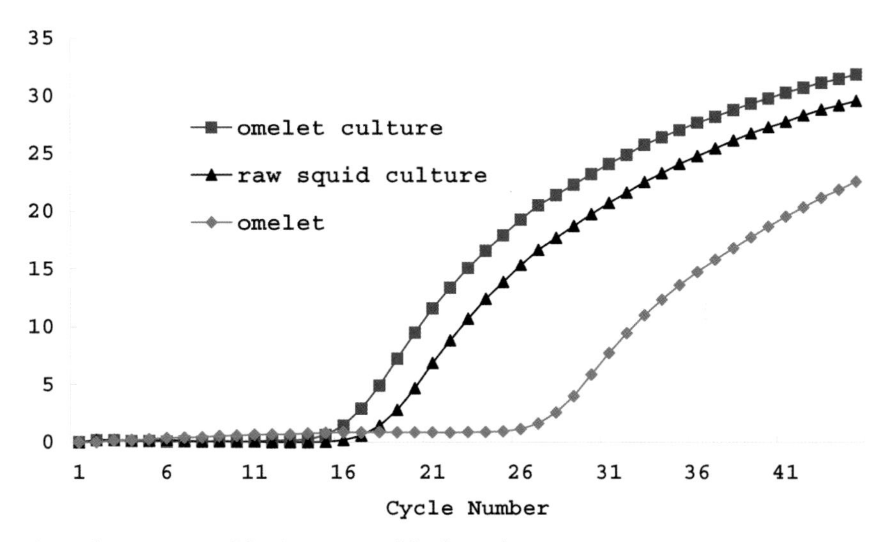

Fig. 3. Fluorescent amplification curves of food samples

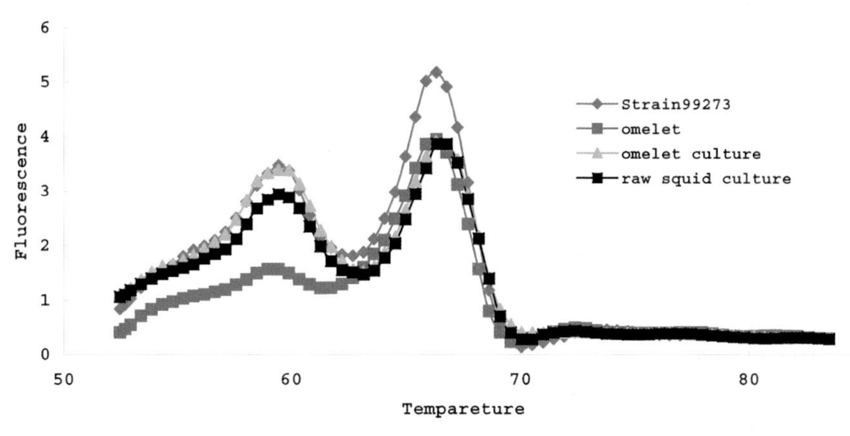

Fig. 4. Melting curves obtained from a human isolate and food samples

Comments

TDH-producing *V. parahaemolyticus* originating in patients with food poisoning is rarely detected in nature or in foods that cause food poisoning. This is because TDH-producing *V. parahaemolyticus* in the environment is usually viable, but nonculturable (VNC). From 1967 to 1997 in our institute, we only detected seven cases of TDH or TDH-producing *V. parahaemolyticus* from suspect foods that had the same serotype as the strain found in the feces of patients. Using culture and the real-time PCR method described here, we identified three new cases in the 3 years from 1998 to 2000 that had matching serotypes. Our method is highly effective in clarifying the distribution of TDH-producing *V. parahaemolyticus* in nature and foodstuffs. In the melting curves shown in Fig. 4, the peaks at 59°C may be attributable to TDH1 reported by Nishibuchi et al. [7].

References

1. Shirai H, Ito T, Hirayama T, Nakabayashi Y, Kumagai K, Takeda Y, Nishibuchi M (1990) Molecular epidemiologic evidence for association of thermostable direct hemolysin (TDH) and TDH-related hemolysin of *Vibrio parahaemolyticus* with gastroenteritis. Infect Immun 58:3568–3573
2. Kishihita M, Matsuoka N, Kumagai K, Yamasaki S, Takeda Y, Nishibuchi M (1992) Sequence variation in the thermostable direct hemolysin-related hemolysin (*trh*) gene of *Vibrio parahaemolyticus*. Appl Environ Microbiol 58:2449–2457
3. Honda T, Ni YX, and Miwatani T (1988) Purification and characterization of a hemolysin produced by a clinical isolate of Kanagawa phenomenon-negative *Vibrio parahaemolyticus* and related to the thermostable direct hemolysin. Infect Immun 56:961–965
4. Yamamoto H (2000) Viable but nonculturable state as a general phenomenon of non-sporeforming bacteria, and its modeling. J Infect Chemother 6:112–114
5. Scott AR, Rice DM, Staffan K (2000) *Vibrio vulnificus*: a physiological and genetic approach to the viable but nonculturable response. J Infect Chemother 6:115–120
6. Nishibuchi M, Takeda Y, Tada J, Oohashi T, Nishimura N, Ozaki H, Fukushima S (1992) Methods to detect the thermostable direct hemolysin gene and a related hemolysin gene of *Vibrio parahaemolyticus*. Nippon Rinsho 642:348–352 (in Japnese)
7. Nishibuchi M, and Kaper JB (1990) Duplication and variation of the thermostable direct haemolysin (*tdh*) gene in *Vibrio parahaemolyticus*. Mol Microbiol 4:87–99

Detection of Fungal Pathogens II

Quantification and Speciation of Fungal DNA in Clinical Specimens Using the LightCycler Instrument 179
JUERGEN LOEFFLER, HOLGER HEBART, NORBERT HENKE, KATHRIN SCHMIDT, HERMANN EINSELE

Quantification and Speciation of Fungal DNA in Clinical Specimens Using the LightCycler Instrument

JUERGEN LOEFFLER*, HOLGER HEBART, NORBERT HENKE, KATHRIN SCHMIDT, HERMANN EINSELE

Introduction

Invasive fungal infections have become a major cause of morbidity and mortality in immunocompromised patients such as bone marrow and solid organ transplant recipients, patients receiving intense chemotherapy, AIDS patients, patients with cystic fibrosis, neonates, and severe burn patients [1]. Invasive aspergillosis has become a leading cause of death, mainly among hematology patients. The incidence is estimated to be 25% and the mortality is up to 90% in patients with acute leukemia [2]. Early diagnosis is essential for appropriate and successful antifungal therapy. Conventional tests for the detection of *Aspergillus* spp., such as blood culture and serology, have limited sensitivity and specificity. Diagnostic assays based on in vitro amplification and speciation of fungal DNA show promising sensitivity and specificity, indicating potential for the early diagnosis of invasive fungal infections [3–5]. However, published assays require 6–8 h for fungal DNA extraction, followed by a minimum of 9 h for DNA amplification and amplicon detection. Post-PCR analysis is mainly based on qualitative or semiquantitative methods such as gel electrophoresis or hybridization by Southern blot or PCR-ELISA techniques.

Here we present a quantitative PCR assay for the detection of DNA of the clinically most relevant fungal species, *Aspergillus fumigatus* in blood and other clinical specimens.

Materials

Equipment and Reagents used for the LightCycler PCR

LightCycler instrument (Roche Diagnostics, Mannheim, Germany)
GelStar nucleic acid stain (FMC Bioproducts, Hessisch Oldendorf, Germany)
Sodium chloride, Magnesium chloride, Tris (Merck, Darmstadt, Germany)
Lyticase, recombinant (Sigma, Deissenhofen, Germany)
Amplification Primers (Roth, Karlsruhe, Germany)
Hybridization Probes (TIB MOLBIOL, Berlin, Germany)

* Juergen Loeffler (✉) (e-mail: juergen.loeffler@med.uni-tuebingen.de)
University of Tuebingen, Medizinische Klinik, Abteilung II, Otfried-Mueller-Strasse 10, 72076 Tuebingen, Germany

Proteinase K (Roche Molecular Biochemicals, Mannheim, Germany)
LightCycler-DNA Master SYBR Green I (Roche Diagnostics, Mannheim, Germany)
LightCycler-FastStart Master Hybridization Probes (Roche Diagnostics, Mannheim, Germany)
QIAmp Tissue Kit (Qiagen, Hilden, Germany)

Procedure

Sample Preparation

Fungi were obtained from the German Collection of Microorganisms (DSM) and were cultured on Sabouraud-Glucose-Agar for 72 h at 30°C. Serial dilutions of conidia were prepared with sterile saline suspensions and adjusted to McFarland 0.5 (10^6 conidia/ml). For sensitivity testing, blood samples from healthy volunteers were used. We spiked 10 ml of EDTA-anticoagulated blood with *Aspergillus fumigatus* conidia, i.e., 10^5–10^1 colony forming units (CFU), or previously extracted *Aspergillus fumigatus* DNA (1 ng to 100 fg). For specificity testing, cells from the following moulds were suspended in sterile saline: *Penicillium brevicompactum*, *Penicillium chrysogenum*, *Absidia corymbifera*, *Paecilomyces variotii*, *Alternaria alternata*, *Scopulariopsis brevicaulis*, *Curvularia inaequalis*, *Fusarium* spp., *Rhizopus oryzae*, *Acremonium chrysogenum* or *Aspergillus fumigatus* [3]. For reproducibility testing, amplification of serial dilutions were repeated ten times under identical conditions. Samples containing DNA from human fibroblasts were prepared concurrently as negative controls.
DNA extraction was always performed immediately after sample preparation.

DNA Extraction Method

For erythrocyte lysis, 5 ml of EDTA-anticoagulated blood was incubated in 15 ml hypotonic Red Cell Lysis Buffer (10 mM Tris, 5 mM magnesium chloride, 10 mM sodium chloride), followed by lysis of the leukocytes in 1 ml White Cell Lysis buffer (10 mM Tris, 10 mM EDTA, 50 mM sodium chloride, 0.2% sodium dodecylsulfate, 200 µg/ml proteinase K). Samples were then incubated in 500 µl buffer containing recombinant lyticase for 45 min at 37°C to generate sphaeroblasts (1 U/100 µl, 50 mM Tris, 1 mM EDTA, 0.2% β-mercaptoethanol). Sphaeroblast lysis and protein precipitation was carried out using the QIAmp Tissue Kit (Qiagen) according to the protocol of the manufacturer. DNA was resuspended in 100 µl elution buffer. PCR was performed immediately or extracted DNA was stored at –80°C.

Primer Design

Primers that bind to regions of the 18S rRNA gene which are highly conserved throughout the whole kingdom of fungi enable the detection of all medically relevant fungal pathogens [3]. Co-amplification of human, protozoal, bacterial and viral DNA was excluded. Amplicons were between 490 bp and 504 bp, depending on the fungal species [3] and included several highly variable regions which were used for speciation by specific hybridization probes. Hybridization probes were established for the detection of the clinically most relevant fungal pathogen, *Aspergillus fumigatus* (see PCR protocol for hybridization probes).

SYBR Green I Master Mix for each 20-µl reaction:

	Volume [µl]	[Final]
LightCycler-DNA Master SYBR Green I	2	1X
$MgCl_2$ (25 mM)	1.6	3 mM
Primers (12 uM each)	1	0.6 µM each
H_2O (PCR grade)	5.4	
Total volume	10	

Hybridization Probe Master Mix using the Hot Start technique for each 20-µl reaction:

	Volume [µl]	[Final]
LightCycler FastStart DNA Master Hybridization Probes	2	1X
$MgCl_2$ (25 mM)	1.6	3 mM
Primers (12 µM each)	1	0.6 µM each
Probes (30 µM each)	2	3 µM each
H_2O (PCR grade)	3.4	
Total volume	10	

We added 10 µl of master mix and 10 µl of DNA template to each capillary. Sealed capillaries were centrifuged in a microcentrifuge and placed into the LightCycler rotor.

The following PCR protocol was used for SYBR Green I detection:

- Denaturation for 2 min at 95 °C
- Amplification

Parameter	Value		
Cycles	45		
Type	Quantification		
	Segment 1	Segment 2	Segment 3
Target temperature [°C]	95	62	72
Incubation time [s]	0	10	15
Temperature transition rate [°C/s]	20	20	20
Acquisition mode	None	None	Single
Gains	F1=5		

Table 1. Oligonucleotides

Aspergillus fumigatus (GenBank Accession #M60300)				
	Position	Length	GC (%)	T_m (°C)
Primers				
ATTGGAGGGCAAGTCTGGTG	543	20	55	65.5
CCGATCCCTAGTCGGCATAG	1046R	20	60	64.5
Product	543–1046	504		
Hybridization probes				
LCRed640-TGAGGTTCCCCAGAAGGAAAGGTCCAGC-P	681	28	57.1	72.8
GTTCCCCCCACAGCCAGTGAAGGC-F	711	24	66.7	73.3

- Melting Curve Analysis

Parameter	Value		
Cycles	1		
Type	Melting curve analysis		
	Segment 1	Segment 2	Segment 3
Target temperature [°C]	95	50	95
Incubation time [s]	20	20	0
Temperature transition rate [°C/s]	20	20	0.2
Acquisition mode	None	None	Continuous

For experiments with hybridization probes, the amplification program was modified as follows:

- Denaturation for 9 min at 95°C
- Amplification

Parameter	Value		
Cycles	45		
Type	Quantification		
	Segment 1	Segment 2	Segment 3
Target temperature [°C]	95	54	72
Incubation time [s]	1	15	25
Temperature transition rate [°C/s]	20	20	20
Acquisition mode	None	Single	None
Gains	F1=1, F2=15		

Gel Analysis of PCR Product

PCR products amplified in the SYBR Green I format were centrifuged by reverse centrifugation for 1 min at 100 *g* in Eppendorf tubes. We separated 10 µl aliquots of each amplification product electrophoretically in a 2% TAE agarose gel, followed by staining with GelStar nucleic acid stain (1:10000).

Results

In Vitro Sensitivity and Linear Range

Titration of genomic *Aspergillus* DNA showed an in vitro sensitivity of 100 fg of fungal DNA, corresponding to 10 CFUs. By adding conidia to blood from healthy donors, we achieved a detection limit of 10 CFU/ml of blood. This sensitivity corresponds to the sensitivity of a PCR assay performed in a conventional thermoblock, followed by hybridization with biotin- or digoxigenin-labeled oligonucleotides [6]. The linear range of the assay was from 10^1 to 10^4 *Aspergillus* conidia (Fig. 1).

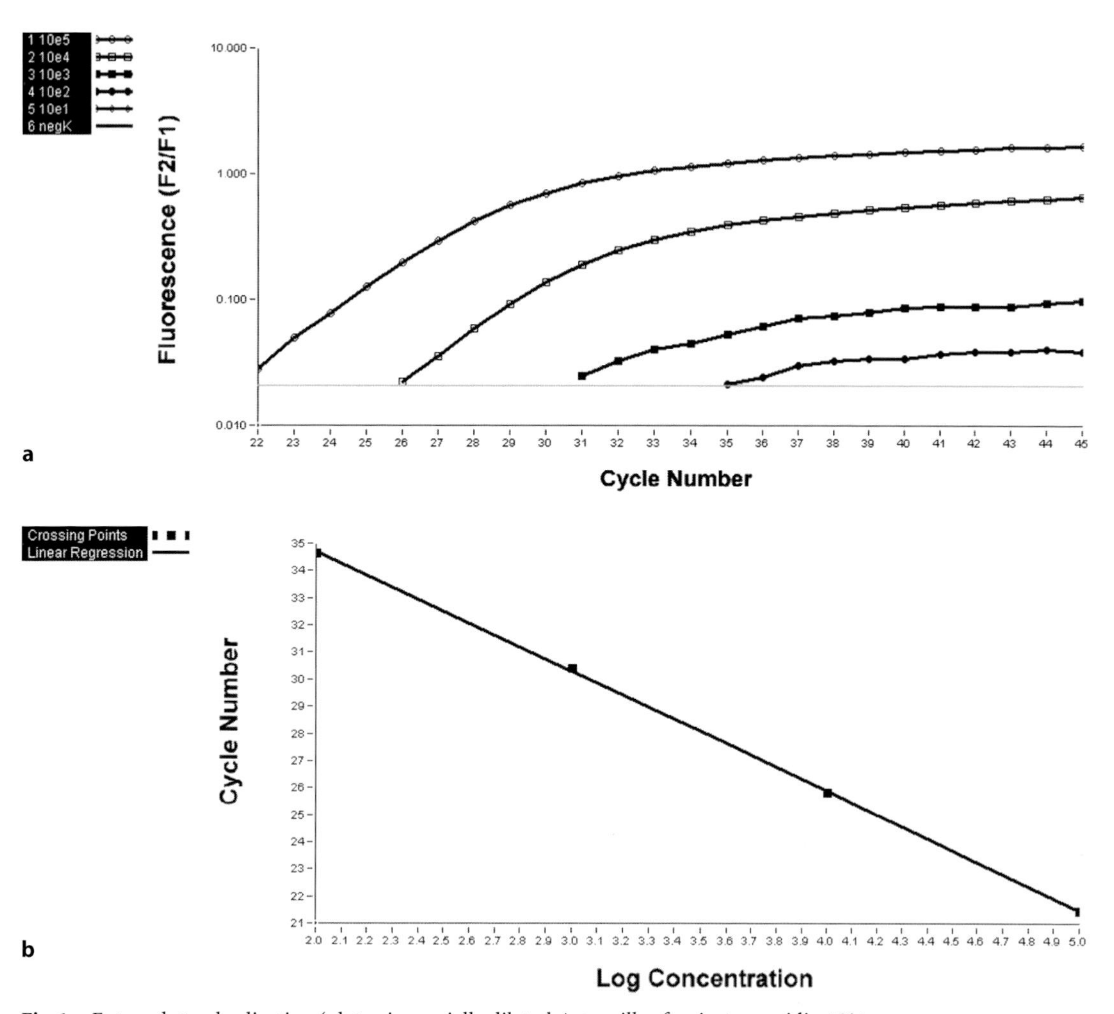

Fig. 1. a External standardization (plot using serially diluted *Aspergillus fumigatus* conidia, 10^4 to 10^1 CFU/ml blood, spiked to blood from healthy donors). The assay was performed with *Aspergillus*-specific hybridization probes. **b** A linear range from 10^1 to 10^4 CFUs was demonstrated which includes the relevant amounts of CFUs found in clinical specimens

Specificity

Hybridization probes designed for *A. fumigatus* hybridized to DNA extracted from *A. fumigatus* cultures. In addition, DNA from the following clinically relevant moulds could be detected: *A. corymbifera*, *P. variotii*, *P. brevicompactum* and *P. chrysogenum*. By melting curve analysis, DNA from cultures of *A. fumigatus*, *P. brevicompactum* and P. *chrysogenum* show an identical probe melting temperature at 72°C, whereas the melting temperature of DNA from *A. corymbifera* is 4°C lower and of DNA from *P. variotii* 9°C lower. The probe sequence of *A. fumigatus*, *P. brevicompactum* and *P. chrysogenum* show 100% identity. DNA extracted from the moulds *A. alternata*, *S. brevicaulis*, *C. inaequalis*, *Fusarium* spp., *R. oryzae* and *A. chrysogenum* as well as the negative control consisting of water remain negative after hybridization with the *A. fumigatus* oligonucleotide (Fig. 2).

Primer Selection and Primer Dimer Formation

Important criteria for primer selection include high specificity (generation of a single amplicon), the ability to detect all clinically relevant fungal species, and a sensitive detection limit with standard amplification kinetics (slope and plateau should be identical among several specific parameters).

Weak primer dimer formation was observed when using the SYBR Green I detection only in samples containing low amounts of fungal DNA (10^1 CFUs). Accumulation of primer dimers may lead to a decrease in sensitivity. For clinical specimens, we strongly recommend using the FastStart DNA Master Hybridization Probes with a hot start PCR protocol. This allows a high throughput of clinical specimens without a significant loss of sensitivity.

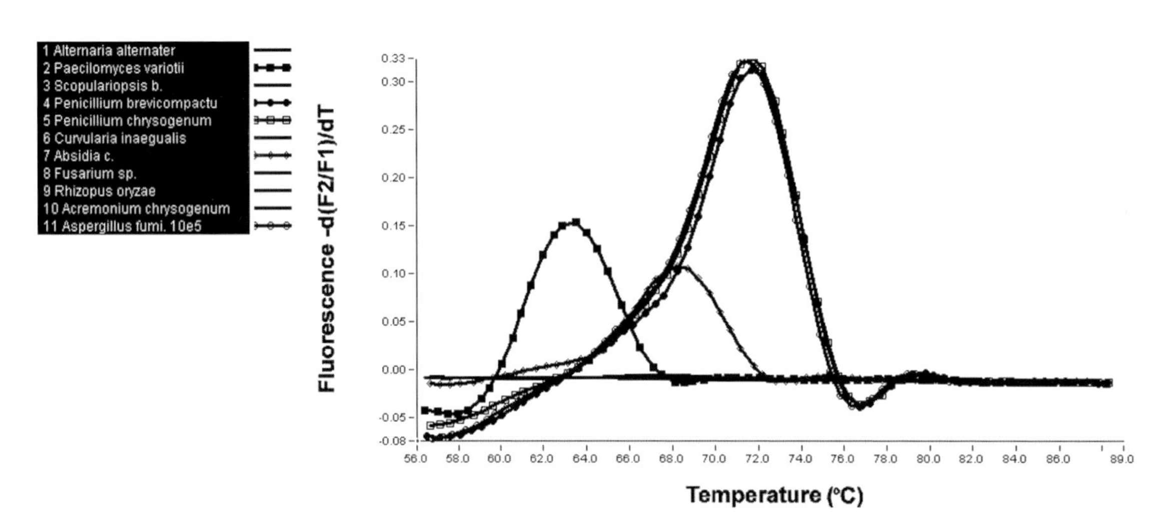

Fig. 2. Specificity of the LC assay. *Aspergillus*-specific hybridization probe detection of *Aspergillus fumigatus* DNA and cross-reaction to *Penicillium brevicompactum* and *Penicillium chrysogenum*. *Absidia corymbifera* DNA showed a 4°C-lower probe melting temperature, *Paecilomyces variotii* DNA showed a 9°C-lower probe melting temperature. No positive signal could be obtained with DNA from *Alternaria alternata*, *Scopulariopsis brevicaulis*, *Curvularia inaequalis*, *Fusarium* spp., *Rhizopus oryzae* and *Acremonium chrysogenum*

Comments

DNA Extraction

Isolation of pure fungal DNA is essential for sensitive detection. The extraction procedure, especially if performed on a routine basis, needs to be rapid and highly reproducible. To prevent false-negative results, especially when extracting DNA from clinical specimens such as blood, serum, bone marrow, tissue and bronchoalveolar lavage, the presence of polymerase inhibitors has to be excluded. Thus, an efficient microformat based on spin columns is recommended. Sensitive and reliable results could be obtained by using the QIAmp Tissue Kit (Qiagen) or the PCR Template Preparation Kit from Roche Molecular Biochemicals.

To monitor for contamination during the DNA extraction procedure (fungal spores), it is necessary to extract concurrently aliquots of sterile saline or blood from healthy volunteers. We recommend that one negative control be extracted for every ten samples analyzed.

Sensitivity of DNA Extraction

Preliminary data indicate that even in patients with histologically proven invasive aspergillosis, only a low number of fungal cells is present in the blood. In contrast, during some viral infections, up to 10^8 copies/ml blood are detectable (Epstein-Barr Virus-associated lymphoproliferative diseases [7]). For external standardization, we used the following clinically relevant amounts of fungal DNA: 10 fg DNA (1 CFU), 100 fg DNA (10^1 CFUs), 1 pg DNA (10^2 CFUs), 10 pg DNA (10^3 CFUs), and 100 pg DNA (10^4 CFUs). Linearity of the assay could be demonstrated within this range.

Specificity of DNA Extraction

Primers that bind to highly conserved gene regions ensure the amplification of fungal DNA even from species which rarely become invasive (e.g. *P. variotii*) or are endemic in certain geographical regions (e.g. *Penicillium* spp.).

Additionally, hybridization probes allow a sensitive and species-specific identification of the clinically most relevant fungal pathogen, *A. fumigatus*, with cross-reaction only to *Penicillium* spp. The species identification is essential for an appropriate, early antifungal therapy. *Aspergillus* spp. and *Penicillium* spp. are treated with amphotericin B. This drug shows high nephrotoxicity whereas other fungal infections might be treated with antifungal drugs showing less severe side effects (e.g., azoles).

Primer Design

For successful and reproducible quantification, optimal primer design is essential. Primer sequences from a conventional PCR assay published previously [3] were transferred to the LightCycler format. For further specificity enhancement, a high annealing temperature (62°C) was chosen.

For optimal sensitivity, PCR diagnosis of fungi should be based on amplifying a multi-copy gene. More than 90 copies of ribosomal genes (18S rRNA gene) are expressed in each fungal cell.

As rapid cycling protocols do not allow generating amplicons larger than 600 bp and the 18S rRNA gene in fungi shows variable regions with a length of 100–150 bp flanked by conserved regions which are used as primer binding sites, we prefer an amplicon with a length of 500 bp.

Storage of the Samples

In order to prevent a loss of sensitivity during storage, extracted DNA samples were kept at -80°C. This is particularly important for DNA extracted from clinical samples, where only a small amount of target DNA might be present.

References

1. Denning D (1998) Invasive aspergillosis. Clin Infect Dis 26:781–805
2. Latgé JP (1999) *Aspergillus fumigatus* and aspergillosis. Clin Microbiol Rev 12:310–350
3. Einsele H, Hebart H, Roller G, Löffler J, Rothenhöfer I, Müller, CA, Bowden RA, van Burik JA, Engelhard D, Kanz L, Schumacher U (1997) Detection and identification of fungal pathogens in blood by using molecular probes. J Clin Microbiol 35:1353–1360
4. Kawamura S, Maesaki S, Noda T, Hirakata Y, Tomono K, Tashiro T, Kohno S (1999) Comparison between PCR and detection of antigen in sera for diagnosis of pulmonary aspergillosis. J Clin Microbiol 37:218–220
5. Van Burik J, Myerson D, Schreckhise R, Bowden R (1998) Panfungal PCR assay for detection of fungal infection in human blood specimens. J Clin Microbiol 36:1169–1175
6. Löffler J, Hebart H, Sepe S, Schumacher U, Klingebiel T, Einsele H (1998) Detection of PCR-amplified fungal DNA by using a PCR-ELISA system. Med Mycol 36:275–279
7. Yamamoto M, Kimura H, Hironaka T, Hirai K, Hasegawa S, Kuzushima K, Shibata M, Morishima T (1995) Detection and quantification of virus DNA in plasma of patients with Epstein Barr Virus-associated diseases. J Clin Microbiol 33:1765–1768

Virology III

Development, Implementation, and Optimization of LightCycler PCR Assays for Detection of Herpes Simplex Virus and Varicella-Zoster Virus from Clinical Specimens . . . 189
Thomas F. Smith, Mark J. Espy, Arlo D. Wold

Qualitative Detection of Herpes Simplex Virus DNA in the Routine Diagnostic Laboratory . . . 201
Harald H. Kessler, Gerhard Mühlbauer, Evelyn Stelzl, Egon Marth

Use of Real-Time PCR to Monitor Human Herpesvirus 6 (HHV-6) Reactivation . . . 211
Junko H. Ohyashiki, Kazuma Ohyashiki, Kohtaro Yamamoto

Simultaneous Identification of Five HCV Genotypes in a Single Reaction . . . 217
Olfert Landt, Ulrich Lass, Heinz-Hubert Feucht

Rapid Detection of West Nile Virus . . . 227
Olfert Landt, Jasmin Dehnhardt, Andreas Nitsche, Gary Milburn, Shawn D. Carver

Use of Rapid Cycle Real-Time PCR for the Detection of Rabies Virus . . . 235
Lorraine McElhinney, Jason Sawyer, Christopher J. Finnegan, Jemma Smith, Anthony R. Fooks

Development of One-Step RT-PCR for the Detection of an RNA Virus, Dengue, on the Capillary Real-Time Thermocycler . . . 241
Boon-Huan Tan, E'ein See, Elizabeth Lim, Eric Peng-Huat Yap

Development, Implementation, and Optimization of LightCycler PCR Assays for Detection of Herpes Simplex Virus and Varicella-Zoster Virus from Clinical Specimens

Thomas F. Smith*, Mark J. Espy, Arlo D. Wold

Introduction

For over 35 years, viruses from clinical specimens have been detected by cell culture techniques; this basic method has been the "gold standard" for the laboratory diagnosis of virus infections [1–4]. Conventionally, specimens were inoculated into cell cultures seeded in tubes; many groups of viruses were recognized by characteristic cytopathic effects induced generally after several days of incubation. In the early 1980s, the incubation time for detection of viruses was significantly reduced by introduction of the shell vial cell culture assay in which early antigens of the virus were detected with monoclonal antibodies after centrifugation and incubation of the vessels for 1–3 days [5, 6].

Description and subsequent application of the polymerase chain reaction (PCR) in the clinical laboratory offered the potential for both rapid and expanded detection of viruses which replicate slowly or not at all in the cell cultures [7, 8]. Nevertheless, concerns of specimen-to-specimen contamination with amplified target nucleic acid restricted implementation of these assays to those laboratories which were specifically constructed to achieve the necessary precautions to prevent false-positive results. Thus, although PCR was more rapid and sensitive than cell-culture-based assays, drawbacks of this demanding technology for resolution of amplified product detection such as gel electrophoresis, Southern blotting, and enzyme-linked immunoassay-based methods were cumbersome for routine laboratory use.

Recently, a major technology breakthrough has resulted in the development of the LightCycler instrument (Roche Molecular Biochemicals, Indianapolis, IN) in which amplified target nucleic acid can be automatically detected by fluorescence after each cycle (denaturation, annealing, and extension). This instrument provides rapid (30–40 min) automation of PCR by precise air-controlled temperature cycling and capillary cuvettes; the continuous monitoring of amplicon development after the annealing step is based on fluorescence resonance energy transfer (FRET) principle. Most importantly, the PCR reaction occurs in a closed system, i.e., the reaction vessels are never opened after the cycling process has started.

* Thomas F. Smith (✉) (e-mail: tfsmith@mayo.edu)
Division of Clinical Microbiology, Mayo Clinic, Hilton 470, 200 First Street S. W., Rochester, MN 55905, USA

This feature, in particular, has allowed use of this instrument for the performance of routine clinical laboratory tests at the Mayo Clinic. We present data for the development, implementation, and optimization of LightCycler assays for the routine laboratory diagnosis of herpes simplex virus (HSV) and varicella-zoster virus (VZV) infections [9–12].

Development of Assay

Herpes Simplex Virus

Primers were directed to target HSV DNA in the polymerase gene and produced an amplified product of 215 bp (Oligo Software, Molecular Biology Insights, Inc., Cascade, CO). In this assay, we obtained a direct relationship of copy number of HSV DNA and the cycle number of detection of the amplicon [9]. For example, 10^9 genomic copies were detected by FRET analysis at cycle number 11; 10 copies were detected at cycle 37.

Using the melting curve feature of the LightCycler software, the two genotypes of the virus could be identified. LightCycler hybridization probes were designed to be homologous to the nucleotide sequence of HSV-2 [melting temperature (T_m); 74.7°C]; in contrast, HSV-1 genotypes differed from HSV-2 in two nucleotide positions which resulted in a T_m of 66.7°C [9].

Varicella-Zoster Virus

Two sets of primers and probes were designed by Oligo software to detect VZV DNA in clinical samples [11]. One set was directed to gene 28 (DNA polymerase); the other primers (gene 29) detected target nucleic acid in the single stranded (SS) DNA binding protein coding area. Both PCR assays detected ≥20 genomic copies of VZV.

Implementation of Assays

Herpes Simplex Virus

Preliminary Evaluation

The technically formatted LightCycler assay was evaluated for the ability to detect HSV DNA from 200 clinical specimens (genital, n=160; dermal, n=38; ocular, n=2) [9]. Specimens extracts were inoculated into shell vial cell cultures and an equal volume of specimen was processed for LightCycler PCR. Of the 200 specimens, 88 (44%) positive results, 69 were detected by both LightCycler PCR and in shell vials (Table 1). Nineteen specimens were detected exclusively by LightCycler PCR; there were no instances in which the shell vial was positive and the LightCycler test was negative (specificity, 100%). Further, specimens positive for HSV by both assays (69) were detected after 26 cycles of LightCycler amplification compared to 33 cycles for specimens (19) positive exclusively by PCR [9]. These data provide strong support of the specificity of the LightCycler assay and the ability of this technology to detect higher copy levels of HSV DNA in samples positive also in cell cultures compared to specimens in which positive results were obtained only by PCR. Importantly, all 19 discrepant LightCycler PCR results were confirmed as positive by another PCR assay directed to the thymidine kinase gene.

Final Evaluation

Our preliminary evaluation of the LightCycler test was performed by (MJE), a research technologist who developed the assay. Prior to offering the LightCycler assay as a replacement for the shell vial cell culture method for routine use in the laboratory, we wanted to expand our experience with this automated PCR system by processing an additional 500 specimens and in addition, to involve other technologists in the laboratory to perform the DNA extraction step and the technical and analytical steps involved in the amplification analysis [10].

Results paralleled those obtained in our preliminary evaluation in that of 500 specimens, 225 (45%) were positive for HSV DNA by LightCycler PCR; 150 (66.7%) of these were also positive by the shell vial assay (Table 2). To resolve the discrepant 75 results (positive by LightCycler PCR but negative by the shell vial assay), all positive samples were sorted according to the cycle number in which amplified HSV DNA was detected. Of the original 225 LightCycler positive samples, 158 were detected within the first 30 cycles of amplification. All of these samples were resolved as specific positive results by recovery of the virus in cell cultures (135) or by PCR directed to the HSV thymidine kinase gene (23).

Of the remaining 67 samples (225 minus 158) detected after cycle 30, 39 specimens were confirmed as true-positive results by detection with the thymidine kinase PCR assay. Therefore, of the original 225 LightCycler positive samples, 197 [158 (≤30 cycles) plus 39 (>30 cycles)] were confirmed by a second test specific for the presence of HSV. Interestingly, the 28 unconfirmed samples (225 minus 197) likely represented samples that were actually positive for HSV DNA but contained low copy numbers of the virus nucleic acid since they were detected late in

Table 1. Detection of herpes simplex virus DNA by LightCycler PCR from clinical specimens (preliminary evaluation)

		Shell Vial Cell Culture		
		Pos	Neg	Total
LightCycler (DNA polymerase gene)	Pos	69	19	88
	Neg	0	112	112
	Total	69	131	200
	P≤0.0001			

Table 2. Detection of herpes simplex virus DNA by LightCycler PCR and by recovery of the virus in shell vial cell culture (final evaluation)

		Shell Vial Cell Culture		
		Pos	Neg	Total
LightCycler (DNA polymerase gene)	Pos	150	75	225
	Neg	0	275	275
	Total	150	350	500
	P<0.0001			

the amplification run (cycler numbers >30). In addition, LightCycler primers and probes designed to detect target DNA within the DNA polymerase gene of HSV were more sensitive than the alternative PCR (thymidine kinase) assay (all specimens positive by the thymidine kinase assay were detected also by the PCR assay directed to the DNA polymerase gene, but 18 of the 67 samples were not positive with the thymidine kinase PCR) (Table 3). Finally, in all of our comparisons, we never had a specimen that was exclusively positive by the shell vial assay.

Based on these developmental data, we implemented the LightCycler assay (May 8, 2000) for both processing dermal and genital specimens as a routine procedure and as a replacement for shell vial cell culture for the laboratory diagnosis of HSV infections. Consistent with our earlier findings, specimens detected during the first 30 cycles of amplification were reported with no further testing. Conversely, those specimens detected after cycle number 30 were reported as positive only after confirmation by the second thymidine kinase target PCR assay (Table 3).

Varicella-Zoster Virus

The LightCycler PCR test (gene 28) was validated only for the detection of VZV DNA from dermal specimens. Of 253 total specimens, 44 (17.4%) were positive for VZV DNA [11]. Of these 44, only 23 (52.3%) were also detected by the shell vial assay. LightCycler PCR produced an increased sensitivity of 93% compared with detection of the virus in shell vial cell cultures (Table 4).

All specimens were tested in parallel with a LightCycler assay directed to target sequences of gene 29 of VZV. All 44 specimens, in addition to 6 other samples, were positive for VZV DNA. We implemented the LightCycler PCR test [DNA polymerase (gene 28)] in which all positive results (44) were confirmed by the PCR test directed to an alternate target (SS binding protein, gene 29). There was total agreement with both LightCycler PCR assays for the remaining 203 specimens that were negative for VZV DNA (specificity, 100%).

Table 3. Detection of herpes simplex virus DNA according to cycle number of LightCycler PCR

	PCR	
Cycle number	Pol	TK
31	8	5
32	10	7
33	6	6
34	4	2
35	6	3
36	10	7
37	4	2
38	10	5
39	3	1
40	2	1
41	4	0
Total	67	39

Table 4. Detection of varicella-zoster virus DNA by LightCycler PCR and by recovery of the virus in shell vial cell cultures

		Cell Culture		
		Pos	Neg	Total
LightCycler (DNA polymerase gene)	Pos	23	21	44
	Neg	0	209	209
	Total	23	230	253
	$P \leq 0.0001$			

Optimization of the Assay

Result reporting data were electronically retrieved retrospectively to assess the efficiency of our algorithm for the determination of HSV DNA in dermal and genital specimens. Over a 6-month period of time, we processed 4,621 specimens; the majority (69%) were from genital sources (Table 5). These data were sorted according to our criteria for reporting results, i.e., whether or not HSV DNA was detected before or after 30 cycles of amplification. HSV was detected in 1,535 of 4,621 (33.2%) specimens. Of the 1,535 positive specimens, 986 (64%) were detected during the first 30 cycles of amplification; no further testing was required since these samples were considered true-positive results. However, a total of 549 (35.8%; 485 confirmed, 64 not confirmed) by the thymidine kinase PCR of the 1,535 specimens were positive for HSV DNA ≥30 cycles of amplification, and required supplementary testing with the second PCR assay (thymidine kinase) prior to issuing a final report to the clinical service. Thus, with the current algorithm and reporting format, a second PCR assay was required for 549 (38.5%) of the specimens yielding HSV DNA in order to sort out the 64 (4.2%) of specimens that were ultimately unconfirmed by the second PCR assay.

For additional analysis, the 549 specimens in which HSV DNA was detected between cycles 31 to 40 were sorted according to the specific cycle number in which the specimen produced a positive FRET signal in the LightCycler assay (Table 6). To reduce the number of unnecessary confirmatory PCR tests performed in the LightCycler assay, we arbitrarily considered that every specimen in which HSV DNA was detected by the initial test through cycle number 37 (DNA polymerase gene target) was a true-positive result. With this criterion, we could immediately report 408 (84%) of the specimen results. Similarly, this algorithm would include 18 (28%) of the 64 specimens that were not originally confirmed by the second PCR (thymidine kinase gene target). Based on previous developmental data, these specimens likely contained very low copy levels of HSV DNA but were not positive by the second PCR method for the several reasons noted above. From a workload standpoint, 426 results for HSV DNA (408+18) detected during cycles 31 through 37 (Table 6) in addition to 986 (Table 5) detected during the first 30 cycles of amplification [total, 1,412 of 1,535 (92%)] could have been result-reported with a single LightCycler PCR (DNA polymerase) based on this

Table 5. Laboratory detection of herpes simplex virus DNA by LightCycler PCR from genital and dermal specimens (*n*=4621), May 8–Oct 31, 2000

	Result (% positive)			
	Cycle no., gene target			
	≤30	31–40	31–40	
Month	Polymerase	Polymerase, TK	Polymerase	
May	177	61	15	
June	113	115	2	
July	84	115	11	
August	210	67	22	
September	201	55	7	
October	201	72	7	
Total	986 (64.2)	485 (31.6)[a]	64 (4.2)[a,b]	1,535 (100.0)

[a] 549/1535=35.8% of polymerase positive results required testing with TK primers.
[b] 1471/1535=95.8% polymerase, TK positive; 64/1535=4.2% TK negative.

Table 6. Confirmation of LightCycler PCR (DNA polymerase gene) results by LightCycler PCR (thymidine kinase gene) – genital and dermal specimens, May 8–October 31, 2000

		Cycle no., gene target (%)				
Reports	*LightCycler Assay*	Cycle number	31–40 polymerase, TK		31–40 polymerase	
		31	62 (12.8)		1 (1.6)	
		32	74 (15.3)		0 (0)	
HSV	*DNA polymerase gene*	33	66 (13.7)		0 (0)	
DNA		34	65 (13.4)	Σ=408 (84%)	2 (3)	Σ=18 (28%)
Detected		35	59 (12.2)		4 (6.3)	
		36	49 (10.1)		5 (7.8)	
		37	33 (6.8)		6 (9.3)	
HSV DNA detected	*Thymidine kinase gene*	38	54 (11.1)		32 (50)	
HSV DNA	*Not tested for thymidine kinase gene*	39	15 (3.1)		8 (12.5)	
Not detected		40	8 (1.7)		6 (9.3)	
		Total	485		64	

retrospective analysis. HSV DNA detected in specimens after ≥38 amplification cycles would not be considered as true positive results for HSV DNA and would be reported as a negative result.

Interim Evaluation of Assay Performance

LightCycler PCR was implemented as a routine test to replace the shell vial cell culture assay for the routine laboratory diagnosis of HSV (genital, dermal) and VZV (dermal) infections. A trend analysis was performed in order to compare the performance characteristics of each test method during the identical time periods, but separated by exactly 1 year. Therefore, archival data of reports for the detection of HSV from genital specimens and HSV and VZV from dermal sites using the shell vial cell culture assay during the period May 8, 1999 through May 8, 2000 were electronically retrieved and compared to report results obtained by LightCycler PCR during the period from May 8, 2000 through May 8, 2001.

Genital Specimens Detection Rates

LightCycler PCR produced a 4.1% increase in the rate of detection of HSV compared with the shell vial cell culture assay even though 233 fewer specimens were processed by the nucleic amplification method. Since the numbers of specimens processed during the two time periods were different (shell vial assay, 5,956; LightCycler PCR, 5,723) the number of HSV positive results were standardized for 5,723 specimens (Table 7). Using this criterion, 1,962 HSV isolates would have been recovered from 5,723 specimens; the 2,199 HSV DNA positive results obtained by LightCycler PCR represented a 12% increase in the detection of this virus infection compared with the cell culture method.

Interestingly, differentiation of the two strains of HSV (types 1 and 2) differed by only 4% during the two time periods regardless of the method, i.e., serotype with shell vial cell cultures or genotype by LightCycler PCR.

Result Report Turnaround Times

The results for the detection of HSV DNA from the genital tract by LightCycler PCR were available the same day (0–24 h) for 60% of the specimen received from this specimen source (Fig. 1). The first results were available (and comparable to LightCycler results) only 24–48 h after receipt of the specimen.

Table 7. Comparative detection of HSV DNA by LightCycler PCR and recovery of the virus by shell vial cell culture assay from genital specimens

	Number specimens	No. positive (%)	HSV-1 (%)	HSV-2 (%)
SVA125	5,956	2,042	762	1,281
(5/8/99 to 4/8/00)		(34.3)	(37.3)	(62.7)
LC	5,723	2,199	732	1,467
(5/8/00 to 4/8/01)		(38.4)[a]	(33.3)	(66.7)

[a] 12% increase compared with shell vial cell culture assay.

Dermal Specimens

Detection Rates

HSV. LightCycler PCR produced a 4.1% increase in the rate of detection of HSV representing a 17.2% increase compared with the shell vial cell cultures (Table 8). Differentiation of the two types of HSV varied by only 1.4% (Type 1) and 1.6% (Type 2) by shell vial cell culture and LightCycler PCR.

VZV. As expected, the rate of detection of VZV by LightCycler PCR was 7.4% higher (71.3% increase) than obtained by recovery of the virus in shell vial cell cultures (Table 8).

Table 8. Comparative detection of HSV and VZV DNA by LightCycler PCR and recovery of the virus by shell vial cell culture assay from dermal specimens

	Virus HSV	HSV	Type 1	Type 2	VZV	HSV, VZV
	Number specimens	No. Pos (%)	No. (%)	No. (%)	No. pos (%)	No. pos (%)
SVA (5/8/99 to 4/8/00)	2,530	598 (23.6)	287 (48.0)	311 (52.0)	261 (10.3)	859 (34.0)
LC (5/8/00 to 4/8/01)	2,533	701 (27.7)[a]	327 (46.6)	374 (53.4)	447 (17.7)[b]	1,148 (45.3)

[a] 17.2% increase compared with SVA.
[b] 71.3% increase compared with SVA.

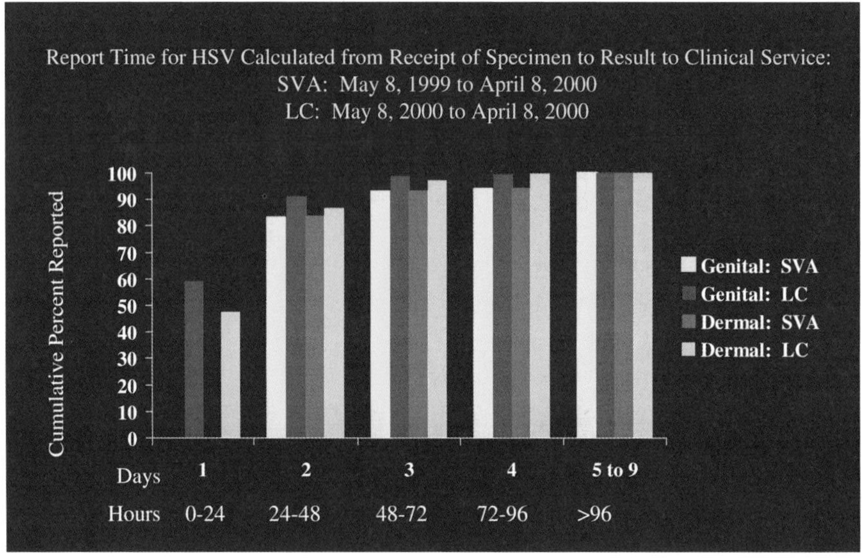

Fig. 1. Report time for HSV calculated from receipt of specimen to remer to clinical service:
SVA: May 8, 1999 to April 8, 2000
LC: May 8, 2000 to April 8, 2000

Result Report Turnaround Times

HSV. Reporting times for dermal specimens (48%) containing HSV DNA were comparable to those from genital sites (60%) with approximately one-half of the reports available to the clinical service the same day on which these specimen types were received (Fig. 1).

VZV. Similar to the turnaround times obtained with HSV DNA, over 55% of the VZV DNA reports were available to attending physicians the same day the dermal specimen was received with the LightCycler PCR assay (Fig. 2). In contrast, the earliest result reports were issued only after 2 days with the shell vial cell culture assay. Importantly, all results with LightCycler PCR (90% by 2 days) were issued to clinical services by day 3. At least 5 days were required for equivalent result reporting based on the shell vial cell culture assay.

Extraction of Nucleic Acids for HSV DNA

Preliminary to the implementation of LightCycler PCR, our laboratory extracted nucleic acids from dermal and genital specimens by the manual IsoQuick (Orca Research, Inc., Bothell, WA). With a current workload of 800 to 900 specimens per month from these clinical sources, an automated instrument for nucleic acid extraction was required as an alternative method to facilitate the processing of specimens preparatory to LightCycler PCR. For this purpose, we compared the

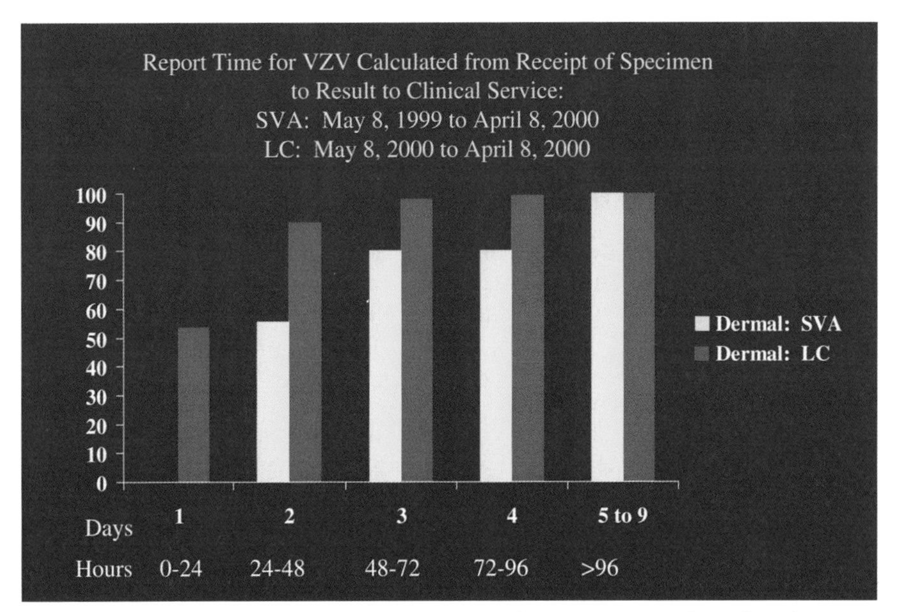

Fig. 2. Report Time for VZV Calculated from Receipt of Specimen to Result to Clinical Service:
SVA: May 8. 1999 to April 8. 2000
LC: May 8. 2000 to April 8. 2000

MagNA Pure instrument as an effective replacement for the manual IsoQuick method with the goal of implementing a cost-efficient, standardized system for the processing of clinical specimens. Of 95 total positive HSV DNA results obtained by the IsoQuick method (3 specimens exclusively detected by this method), 92 were detected after extraction of the specimens by the MagNA Pure instrument (P=NS). These data indicated that the automated MagNA Pure instrument and the IsoQuick method were equivalent for the nucleic acid extraction of specimens preparatory to LightCycler PCR. The automated extraction is anticipated to be especially applicable for use in laboratories processing high numbers of specimens based on its performance in providing standardized, reproducible, and cost-effective DNA extraction preparatory to LightCycler PCR.

Replacement of Cell Culture Methods with LightCycler PCR in Clinical Virology

With implementation of LightCycler PCR assays for the diagnosis of HSV and VZV infections, our laboratory has replaced 77% (HSV, 71%; VZV, 6%) of the diagnostic procedures and 47% of the cell cultures which previously used this type of methodology.

More recently, we have developed a LightCycler PCR assay for the detection of CMV DNA from urine specimens; collectively, with the three molecular assays (HSV, VZV, CMV), we are able to diagnose 85% of the viruses and replace 53% of cell culture requirements by implementation of these tests.

The remaining 15% of the viruses which are recovered by cell culture methods (enteroviruses, adenoviruses, influenza and parainfluenza viruses, respiratory syncytial virus) now require 47% of our cell culture resources.

Glass Cuvette (Capillary) Breakage

One of the potential disadvantages of LightCycler PCR for routine diagnostic use is the use of optical grade glass cuvettes as vessels for sample in the assay. The obvious concern was possible breakage of the cuvettes upon removal of the samples from the carousel after the PCR amplification has taken place and the threat of carryover contamination to other specimens with amplified target DNA possibly leading to false-positive results.

Importantly, a review of our logs of individual test runs revealed breakage of cuvettes on only 6 occasions out of a total of 5,739 specimens (0.1%). Thus, the risk of a potential amplicon contaminating event is minimal; in addition uracil-*N*-glycosylase (UNG) enzyme is added to the master mix for each specimen. When rare breakage events occur, the carousel is immediately decontaminated by reagents containing bleach and detergents.

We feel that the multiple advantages of the closed and contained amplification format and real-time detection characteristics of LightCycler PCR greatly outweigh the minimal probability of cuvette breakage with a subsequent amplicon contamination event especially when compared with conventional methods of nucleic acid amplification and product detection in many laboratories performing molecular analysis for detection of virus nucleic acids.

Conclusion

LightCycler PCR, in trend analysis data, was more sensitive [HSV, 12%(genital) to 17% (dermal); VZV, 71% (dermal)] and had a faster turnaround time [HSV, 59% (genital) to 48% (dermal); VZV, 55% (dermal)] results reported the same day the specimen is received) compared to reports issued at least 1 day later for the shell vial cell culture method. Nucleic acids preparatory to the LightCycler PCR step can be efficiently and reproducibly extracted with the automated MagNA Pure instrument; the instrument is most applicable to processing high numbers of specimens. LightCycler PCR has replaced cell culture assay for 85% of the specimens yielding viruses (HSV, 71%; VZV, 6%; CMV, 8%) and 53% of the cell culture requirements for detection of these viruses from clinical specimens.

We feel that LightCycler PCR creates a new dimension for detection of HSV, VZV, and CMV infections. Our goal is replacement of conventional cell culture methods by PCR and the addition of other qualitative and quantitative assays for noncultivatible viruses.

References

1. Smith TF, Wold AD, Espy MJ, Marshall WF (1993) New developments in the diagnosis of viral diseases. Infect Dis Clin North Am 7:183–201
2. Smith TF, Wold AD, Espy MJ (1992) Diagnostic virology – then and now. Advances in Experimental Med & Biol 312:191–199
3. Storch GA (2000) Diagnostic virology. Clin Infect Dis 31:739–751
4. Hsiung GD (1989) The impact of cell culture sensitivity on rapid viral diagnosis: a historical perspective. Yale J Biol Med 62:79–88
5. Gleaves CA, Smith TF, Shuster EA, Pearson GR (1984) Rapid detection of cytomegalovirus in MRC-5 cells inoculated with urine specimens by using low-speed centrifugation and monoclonal antibody to an early antigen. J Clin Microbiol 19:917–919
6. Shuster EA, Beneke JS, Tegtmeier GE, Pearson GR, Gleaves CA, Wold AD, Smith TF (1985) Monoclonal antibody for rapid laboratory detection of cytomegalovirus infections: characterization and diagnostic application. Mayo Clin Proc 60:577–585
7. Persing DH, Smith TF, Tenover FC, White TJ (eds.) Diagnostic Molecular Microbiology: Principles and Applications (1993) American Society for Microbiology, Washington, DC
8. Espy MJ, Smith TF, Persing DH (1993) Dependence of polymerase chain reaction product inactivation protocols on amplicon length and sequence composition. J Clin Microbiol 31:2361–2365
9. Espy MJ, Uhl JR, Mitchell PS, Thorvilson JN, Svien KA, Wold AD, Smith TF (2000) Diagnosis of herpes simplex virus infections in the clinical laboratory by LightCycler PCR. J Clin Microbiol 38:795–799
10. Espy MJ, Ross TK, Teo R, Svien KA, Wold AD, Uhl JR, Smith TF (2000) Evaluation of LightCycler PCR for implementation of laboratory diagnosis of herpes simplex virus infections. J Clin Microbiol 38:3116–3118

11. Espy MJ, Teo R, Ross TK, Svien KA, Wold AD, Uhl JR, Smith TF (2000) Diagnosis of varicella-zoster virus infections in the clinical laboratory by LightCycler PCR. J Clin Microbiol 38:3187–3189
12. Espy MJ, Rys PN, Wold AD, Uhl JR, Sloan LM, Jenkins GD, Ilstrup DM, Cockerill FR III, Patel R, Rosenblatt JE, Smith TF (2001) Detection of herpes simplex virus DNA in genital and dermal specimens by LightCycler PCR after extraction using the IsoQuick, MagNA Pure, and BioRobot 9604 methods. J Clin Microbiol 39:2233–2236

Qualitative Detection of Herpes Simplex Virus DNA in the Routine Diagnostic Laboratory

Harald H. Kessler*, Gerhard Mühlbauer, Evelyn Stelzl, Egon Marth

Introduction

Herpes simplex virus (HSV) causes a wide spectrum of clinical manifestations in the central nervous system. Neonatal HSV infection following exposure to the virus at delivery can produce severe disseminated infection and death. Effective therapeutic management exists today, however, antiviral drugs must be administered early. Therefore, rapid laboratory diagnosis is essential for decreasing the lethality as well as the sequelae of HSV infection.

Today, PCR is recognized as a reference standard assay for the sensitive and specific detection of HSV DNA [1, 2]. Studies on detection of HSV DNA are usually based on home-brew protocols. These assays commonly involve complicated and time-consuming extraction procedures and pose potential problems of false positive results due to crossover contamination. Therefore, they prove impractical for routine diagnostic laboratories. To meet the needs of these laboratories, simplified and automated methods are preferable.

The MagNA Pure LC (Roche Diagnostics, Mannheim, Germany), a fully automated specimen preparation instrument, and the LightCyler instrument (Roche) have recently been introduced. Combined employment of these instruments allows establishment of rapid molecular assays for detection of pathogens in the routine diagnostic laboratory.

In this article, we describe a qualitative molecular assay for detection of HSV DNA based on sample preparation with the MagNA Pure LC followed by LightCycler-PCR. The performance of this assay was evaluated and clinical specimens were tested.

Materials

Equipment

MagNA Pure LC instrument (Roche Diagnostics, Mannheim, Germany)
LightCycler Carousel Centrifuge (Roche Diagnostics)
LightCycler Instrument (Roche Diagnostics)

* Harald H. Kessler (✉) (e-mail: harald.kessler@uni-graz.at)
Molecular Diagnostics Laboratory, Institute of Hygiene, Karl-Franzens-University, Universitaetsplatz 4, 8010 Graz, Austria

LightCycler Capillaries (Roche Diagnostics)
LightCycler Centrifuge Adaptors (Roche Diagnostics)
LightCycler software vers. 3.01 (Roche Diagnostics; for runs)
LightCycler software vers. 3.5 (Roche Diagnostics; for data analysis)

Kits

MagNA Pure LightCycler Total Nucleic Acid Kit (Roche Diagnostics)
LightCycler Fast Start DNA Master Hybridization Probes (Roche Diagnostics)

Reagents

Amplification Primers (TIB MOLBIOL, Berlin, Germany)
TaqMan Probe (TIB MOLBIOL)

Procedure

Study Design

Initially, the inter-assay variation and the detection limit were tested. The Second European Union Concerted Action HSV Proficiency Panel, which consisted of 12 vials with different concentrations of HSV type 1 (HSV-1), strain MacIntyre (American Type Culture Collection), HSV type 2 (HSV-2), strain MS (American Type Culture Collection), VZV, and negative samples, were used (Table 1). Samples were run three times on different days with the automated molecular assay, which consisted of DNA extraction on the MagNA Pure LightCycler instrument and LightCycler-PCR. Furthermore, results were compared with a molecular assay based on a manual extraction protocol and the identical LightCycler-PCR protocol. This assay has been described in detail elsewhere [3, 4].

Intra-assay variation of the automated molecular assay was then tested. Plasma was collected from a patient without clinical presentations compatible with HSV infection. Aliquots were spiked with tenfold dilutions of a culture supernatant of commercially available HSV-1-infected Vero cells (VR-260; American

Table 1. Second European Union Concerted Action Herpes Simplex Virus Proficiency Panel

Vial number	Virus	Estimated no. of genomes per ml
1	Negative	–
2	HSV-2	$3\text{–}7 \times 10^5$
3	HSV-1	$0.7\text{–}2 \times 10^3$
4	HSV-1	$2\text{–}6 \times 10^7$
5	VZV	–
6	HSV-2	$3\text{–}7 \times 10^2$
7	HSV-1	$2\text{–}6 \times 10^5$
8	HSV-2	$3\text{–}7 \times 10^6$
9	HSV-1	$2\text{–}6 \times 10^2$
10	HSV-2	$3\text{–}7 \times 10^3$
11	HSV-1	$2\text{–}6 \times 10^3$
12	Negative	–

Type Culture Collection, Rockville, Md.). Each dilution was run eight times. Each run contained five positive and three negative controls (water).

Next, precision and influence of different sample materials were tested. Whole blood (3 ml EDTA anticoagulated), serum, and plasma were collected from a patient without clinical presentations compatible with HSV infection. Aliquots were spiked with a culture supernatant of HSV-1 as described above. Different sample volumes were run three times with the automated molecular assay. Each run contained three negative controls (blank reagent and water).

Finally, a total of 76 clinical specimens were investigated. Forty-two cerebrospinal fluid (CSF) samples were obtained from patients who were admitted to the Department of Pediatrics, University Hospital Graz, Austria. All of them had clinical presentations compatible with HSV encephalitis. Thirty-four vaginal swabs were collected from pregnant women at the outpatient clinics of the Department of Obstetrics and Gynecology, University Hospital Graz. All of them had clinical presentations compatible with genital HSV infection.

Sample Preparation

The MagNA Pure LC software, version 2.0 was used. The extraction protocol "Total NA Serum, Plasma, Blood" was employed. For steps one and two, a 200 µl sample volume was used and the sample volume for the third step was either 50, 100, 150 or 200 µl. An elution volume of 100 µl and a dilution volume of 0 µl was chosen in each run. Other details, including reagent volumes and number of reaction tips needed for the run, were automatically calculated by the software. The MagNA Pure LC automatically performed all remaining steps of the procedure with specially designed nuclease-free, disposable reaction tips. These reaction tips not only transferred the samples, but also served as reaction vials for the procedure. Within the tips, nucleic acids were bound to magnetic beads, washed free of impurities, and finally eluted from the magnetic beads into a cooled sample cartridge. During the run, used reaction tips were automatically discarded into an attached, autoclavable waste bag.

After completion of DNA extraction, the MagNA Pure LC Cooling Block, the sample carousel with LC capillaries, and the reaction vessel with master mix were moved into the postelution area. The previously programmed postelution protocol then pipetted 15 µl of the master mix and 5 µl of the processed sample into each of the LC capillaries.

Primer Design

Oligonucleotides deduced from the published sequence of the DNA polymerase gene-coding region from HSV were used [5, 6]. This set of primers, which was chosen within a highly conserved region of the DNA polymerase gene from the herpesvirus group, allows amplification of a 92-bp fragment of the HSV-1 and HSV-2 DNA polymerase genes in clinical samples [7, 8]. The primer and probe sequences and characteristics are shown in Table 2.

Table 2. Oligonucleotides

HSV Type 1 Complete Genome (GenBank Accession #X14112) HSV Type 2 Complete Genome (GenBank Accession #Z86099)				
	Position	Length	GC (%)	T_m (°C)
CATCACCGACCCGGAGAGGGAC	HSV-1, 65866 HSV-2, 66339	22	68.2	71.7
GGGCCAGGCGCTTGTTGGTGTA	HSV-1, 65957 R HSV-2, 66430 R	22	63.6	73.3
Product	HSV-1, 65866–65957 HSV-2, 66339–66430	92		
TaqMan Probe				
FAM-CCGCCGAACTGAGCAGA- CACCCGCGC-TAMRA	HSV-1, 65907 HSV-2, 66380	26	73.1	79.1

LightCycler PCR

TaqMan Master Mix for each 20-μl reaction:

	Volume [μl]	[Final]
LightCycler Fast Start DNA Master Hybridization Probes	2	1X
$MgCl_2$ (25 mM)	2.4	4 mM
Primers (50 μM each)	0.2 + 0.2	0.5 μM each
TaqMan Probe (7.4 μM)	0.5	0.185 μM
H_2O (PCR grade)	9.7	
Total volume	15	

After completion of the postelution protocol, LightCycler capillaries were sealed. Then, the sample carousel with the capillaries was centrifuged in the LightCycler Carousel Centrifuge and placed into the LightCycler. After denaturation for 10 min at 95°C, the amplification protocol was run.

- Amplification

Parameter	Value	
Cycles	55	
Type	Quantification	
	Segment 1	Segment 2
Target temperature [°C]	95	60
Incubation time [s]	2	20
Temperature transition rate [°C/s]	20	10
Acquisition mode	None	Single
Gains	F1=1	

After the final cycle, the capillaries were cooled for 2 s at 40°C. Fluorescence curves were analyzed with the LightCycler software. Cycles 15 to 55 and channel F1 were selected for calculation of crossing points, which were defined as the maximum of the second derivative from the fluorescence curves. Automated calculation was done by the second derivative maximum method.

Results

Inter-assay Variation and Detection Limit

When samples of the Second European Union Concerted Action Herpes Simplex Virus Proficiency Panel were tested with the automated molecular assay, consisting of DNA extraction on the MagNA Pure LC and LightCycler-PCR, 0.7 to $2x10^3$ estimated HSV-1 genomes per ml (vial 3), i.e., 7 to 20 estimated HSV-1 genomes per LightCycler capillary, could consistently be detected. With the dilution containing 2 to $6x10^2$ estimated HSV-1 genomes per ml (vial 9), i.e., 2 to 6 estimated HSV-1 genomes per LightCycler capillary, product was not detected. When HSV-2 samples from the same panel were tested, even the lowest concentration, 3 to $7x10^2$ estimated HSV-2 genomes per ml (vial 6), i.e., 3 to 7 estimated HSV-2 genomes per LightCycler capillary, could consistently be detected. With the LightCycler-PCR assay including the manual extraction protocol, 2 to $6x10^3$ HSV-1 genomes per ml (vial 11), i.e., 20 to 60 estimated HSV-1 genomes per LightCycler capillary could consistently be detected. Both of the less concentrated HSV-1 samples (vials 3 and 9) were below the detection limit. Testing of the HSV-2 samples revealed identical results with the automated molecular assay, i.e., all positive samples of the Second European Union Concerted Action Herpes Simplex Virus Proficiency Panel could be detected. Negative samples (vials 1 and 12) and the VZV sample (vial 5) always gave negative results.

Intra-assay Variation

When reproducibility of the automated molecular assay was tested, crossing points were usually within one cycle. Standard deviations showed a tendency to increase when lower concentration samples were tested (Table 3). Fluorescence curves of repeated samples at three different dilutions are shown in Fig. 1.

Table 3. Reproducibility of the molecular assay, including DNA extraction on the MagNA Pure LC and automated calculation by the second derivative maximum method (LC software, version 3.5).

Estimated no. of HSV-1 genomes per ml	Mean crossing point	Standard deviation
$1x10^6$	32.41	0.22
$5x10^5$	33.57	0.25
$1x10^5$	36.43	0.30
$5x10^4$	37.40	0.22
$1x10^4$	39.38	0.82
$5x10^3$	41.37	0.33
$1x10^3$	43.43	0.62

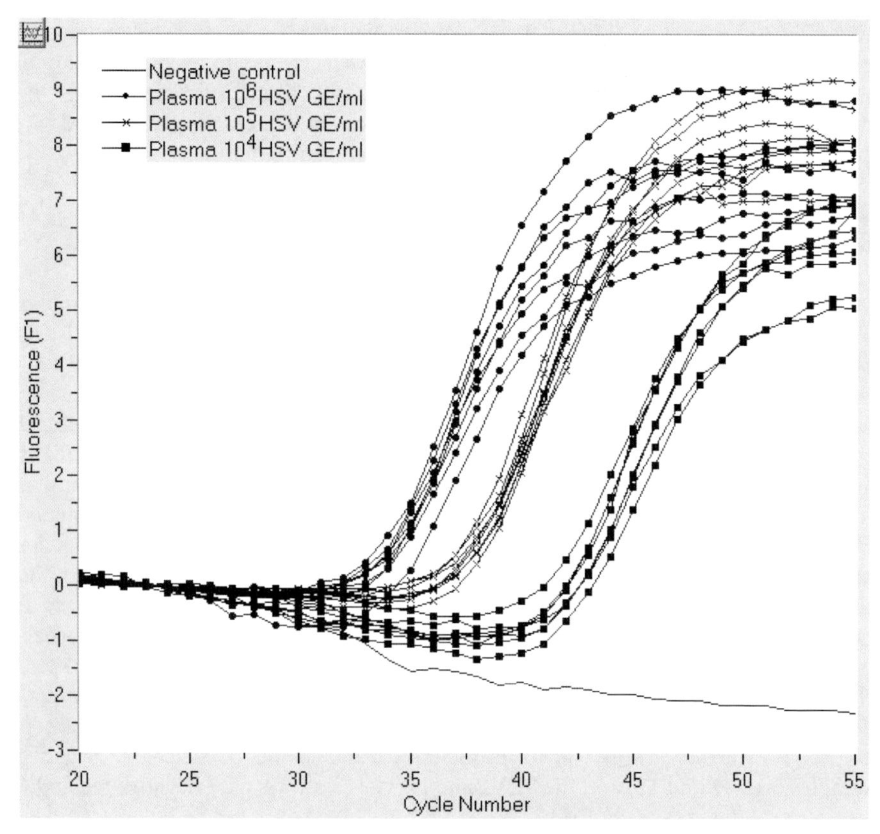

Fig. 1. Screen-shot of reproducibility testing with three different dilutions of plasma which contained an estimated number of 1 x 10^6, 1 x 10^5 or 1 x10^4 HSV-1 genome equivalents per ml

Precision and Influence of Different Sample Materials

When different sample volumes were spiked with a distinct amount of HSV-1 and run on the automated molecular assay, 50, 100, 150, and 200 µl sample volumes were distinguishable by the mean value of triplicates (Table 4; Fig. 2). As shown by the fluorescence curves, the quality of the different DNA extracts was independent from the input sample volume. In addition, different sample materials (EDTA blood, serum, and plasma) showed similar crossing points (data not shown).

Clinical Specimens

When clinical specimens were tested, 2 of 42 CSF samples were found positive, whereas 12 of 34 vaginal swabs gave a positive result. Results of clinical samples (vaginal swabs) are shown in Fig. 3.

The automated DNA extraction with the MagNA Pure LC was completed within 105 min for extraction of 32 samples. This included a 15-min set-up of the MagNA Pure LC. The time required for the postelution protocol was 15 min. After centrifugation, the LightCycler-PCR took another 60 min. No contamination was observed during the entire study.

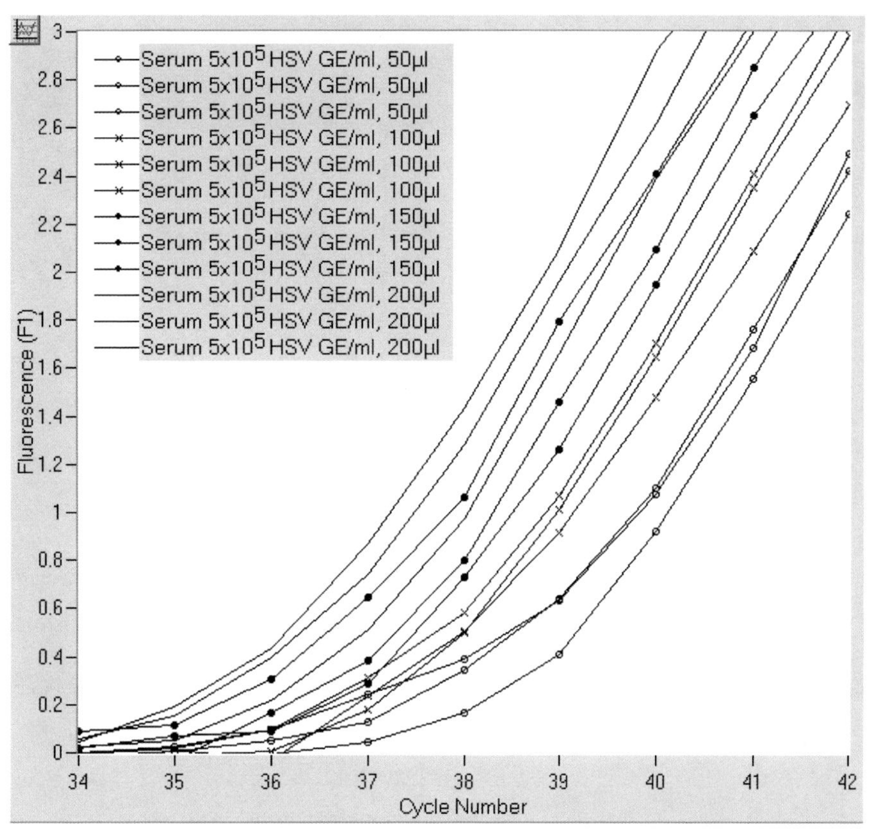

Fig. 2. Screen-shot of testing different amounts of sample volume when testing serum samples which contained an estimated number of 5 x 10^5 HSV-1 genome equivalents per ml

Table 4. Results of different volumes of sera spiked with a distinct amount of HSV-1. Automated calculation (LightCycler software, version 3.5) by the second derivative maximum method.

Amount of serum (µl)	Mean crossing point	Standard deviation
50	37.87	0.22
100	36.56	0.30
150	36.08	0.60
200	35.48	0.23

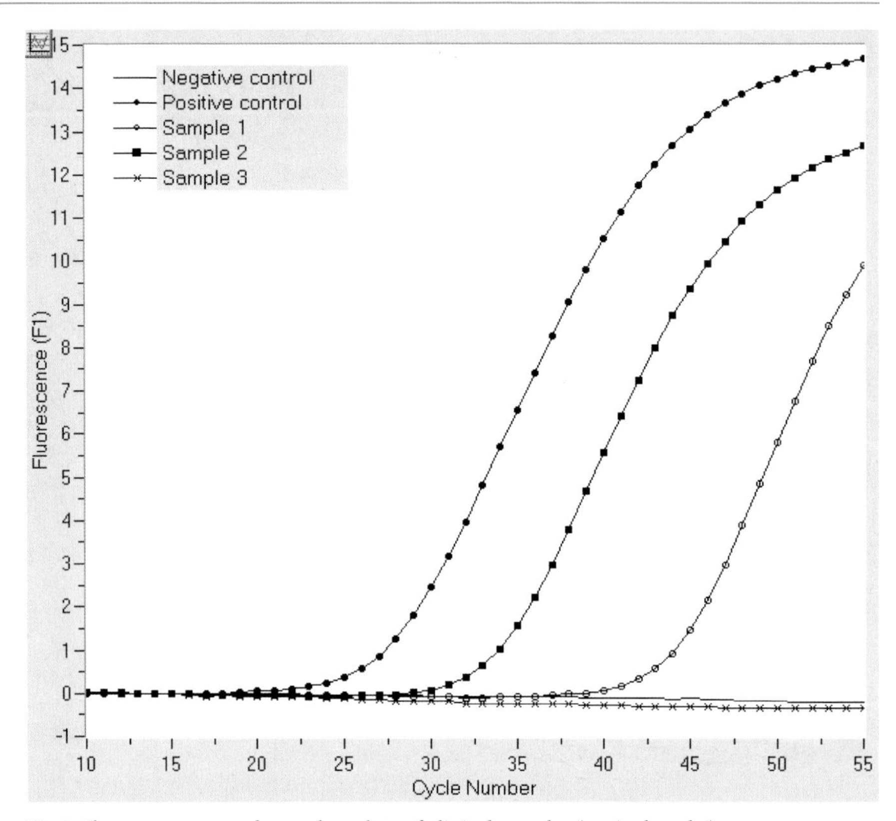

Fig. 3. Fluorescence vs cycle number plots of clinical samples (vaginal swabs)

Comments

Automated Versus Manual Sample Preparation

The detection limit of the molecular assay, which included DNA extraction on the MagNA Pure LC, was close to the theoretical detection limit: 7 to 20 estimated HSV-1 genomes per LightCycler capillary and 3 to 7 estimated HSV-2 genomes per LightCycler capillary were detected. When DNA was extracted with the manual protocol [3, 4], the detection limit for HSV-2 was similar, whereas the detection limit for HSV-1 genomes was slightly higher (20 to 60 estimated HSV-1 genomes per LightCycler run). The difference may be the result of the double sample volume (200 μl compared to 100 μl) used for DNA extraction on the MagNA Pure LC. Furthermore, when the MagNA Pure LC extraction is used, samples are concentrated to a volume of 100 μl, in contrast to the manual protocol, in which the samples are diluted to a volume of 400 μl.

The time for extraction of 32 samples was similar with both extraction methods. If less than eight samples are extracted, the manual extraction may be the faster. However, the hands-on time for extraction is only 15 min when using the MagNA Pure LC.

Use of Automated Molecular Assays in the Routine Diagnostic Laboratory

The automated molecular assay, including DNA extraction on the MagNA Pure LC, was reliable. Repeated testing revealed nearly identical crossing points and no contamination during the entire study. The probability of false-positive results from contamination increases with the number of manual steps involved in sample processing [9, 10]. In this study, no contamination was observed with either extraction protocol. However, this study was done in an ISO 9002-certified laboratory under strict safety precautions. In a routine diagnostic laboratory, exclusion of contamination may be one of the major advantages of a fully automated nucleic acid extraction instrument.

The automated molecular assay proved to be suitable for the routine diagnostic laboratory. Automation of molecular assays helps to avoid human error by increasing precision and reproducibility and providing better standardization.

References

1. Mitchell PS, Espy MJ, Smith TF, Toal DR, Rys PN, Berbari EF, Osmon DR, Persing DH (1997) Laboratory diagnosis of central nervous system infections with herpes simplex virus by PCR performed with cerebrospinal fluid specimens. J Clin Microbiol 35:2873–2877
2. Madhavan HN, Priya K, Anand AR, Therese KL (1999) Detection of herpes simplex virus (HSV) genome using polymerase chain reaction (PCR) in clinical samples. Comparison of PCR with standard laboratory methods for the detection of HSV. J Clin Virol 14:145–151
3. Kessler HH, Mühlbauer G, Rinner B, Stelzl E, Berger A, Dörr HW, Santner B, Marth E, Rabenau H (2000) Detection of herpes simplex virus DNA by real-time PCR. J Clin Microbiol 38:2638–2642
4. Kessler HH (2001) Qualitative detection of herpes simplex virus DNA on the LightCycler. In: S Meuer, C Wittwer, K Nakagawara (eds), Rapid cycle real-time PCR. Springer-Verlag, Heidelberg, pp 331–340
5. Larder BA, Kemp SD, Darby G (1987) Related functional domains in virus DNA polymerases. EMBO J 6: 169–175
6. Tsurumi T, Maeno K, Nishiyama Y (1987) Nucleotide sequence of the DNA polymerase gene of herpes simplex virus type 2 and comparison with the type 1 counterpart. Gene 52: 129–137
7. Cao M, Xiao X, Egbert T, Darragh TM, Yen TSB (1989) Rapid detection of cutaneous herpes simplex virus infection with the polymerase chain reaction. J Invest Dermatol 92: 391–392
8. Brice SL, Krzemien D, Weston WL, Huff JC (1989) Detection of herpes simplex virus DNA in cutaneous lesions of erythema multiforme. J Invest Dermatol 93: 183–187
9. Clewley JP (1989) The polymerase chain reaction, a review of the practical limitations for human immunodeficiency virus diagnosis. J Virol Methods 25:179–188
10. Kwok S, Higuchi R (1989) Avoiding false positives with PCR. Nature (London) 339:237–238

Use of Real-Time PCR to Monitor Human Herpesvirus 6 (HHV-6) Reactivation

Junko H. Ohyashiki*, Kazuma Ohyashiki, Kohtaro Yamamoto

Introduction

The diseases caused by infection with human herpesviruses in immunocompromised patients, particularly in transplant recipients, are well recognized and documented [1]. Human herpesvirus 6 (HHV-6) is classified in the same subfamily with Cytomegalovirus, β herpesvirus. HHV-6 causes a life-long persistent infection in over 90% of people before age 2 years, and is a causative agent of febrile illness in children, particularly exanthem subitum. On the basis of molecular and phenotypic characteristics, HHV-6 isolates are categorized into two groups, variant A (HHV-6A) and variant B (HHV-6B) [2]. HHV-6 is normally controlled by the host T cell immune responses, so HHV-6 disease presents when these cell mediated responses are impaired, either through immunosuppressive drugs or organ transplantation, or through human immunodeficiency virus infection. So far, HHV-6 is known to be associated with encephalitis, fatal interstitial pneumonia and stem cell transplant-related complications such as graft-versus-host disease (GVHD) and delayed engraftment [3].

To date, the most sensitive approach for the detection of HHV-6 has been polymerase chain reaction (PCR). PCR provides a specific technique with which to investigate disease association, however, the conventional PCR method is insufficient to determine clinically relevant viral replication. Since the amount of viral DNA may vary between persistence in normal individuals and during primary infection or episodes of reactivation, and may be higher in immunocompromised patients, quantification of the virus in clinical samples is required.

Several methods for quantitative PCR have been reported, including end-point dilution, quantification using external and internal standards, competitive PCR using a target sequence mimetic, and hybridization of PCR products to probes in a microwell format [2]. The last two methods are considered to be the most accurate and have been applied to many viral detection systems. These tests are, however, labor-intensive and require post-PCR handling. Recently, quantitative, fluorescence-based, real-time PCR assays in closed tube systems have been developed. To establish rapid diagnostic quantification of HHV-6, we developed

* Junko H. Ohyashiki (✉) (e-mail:junko@hh.iij4u.or.jp)
Department of Virology, Medical Research Institute, Tokyo Medical and Dental University, 1-5-45 Yushima, Bunkyo-ku, Tokyo, 113-8510, Japan

a quantitative real-time PCR assay using the LightCycler system and hybridization probes [3].

Materials

Equipment

LightCycler instrument

Reagents

QIAmp Blood kit (Qiagen)
Proteinase K (Roche Diagnostics, Mannheim, Germany)
Amplification primers (Greiner Japan)
Hybridization Probes (Nihon Gene Research Laboratories)
LightCycler-DNA Master Hybridization Probes

Procedure

Standard template DNA: Assays were performed with serial dilutions (10^1–10^5) of a plasmid containing the 101K region of HHV-6. For plasmid construction, the corresponding sequence of the 101K gene region was and inserted into pUC19 Vector [2]. After plasmid preparation, the DNA concentration was determined with a spectrophotometer at 260 nm, and the corresponding copy number was calculated. Serial dilutions of plasmids were prepared in H_2O.

Sample Preparation

DNA from whole blood was extracted with the QIAmp blood kit according to the manufacture's instructions.

Primer and Probe Design

The amplification primers and the hybridization probes used for quantification of HHV-6 are unique based on published sequences where both variant A and B have homology [2]. Therefore, this system quantifies either HHV-6A or HHV-6B. Details are shown in Table 1.

Table 1. Oligonucleotides

HHV-6 (GenBank Accession #X83413)				
	Position	Length	GC (%)	T_m (°C)
Primers				
ACC CGA GAG ATG ATT TTG CG	20,889	20	50.0	62.7
GCA GAA GAC AGC AGC GAG AT	21,100 R	20	60.0	68.1
Hybridization probes				
LCRed640-GGG TCA TTT ATG TTA TAG ACG GT	21,034	23	39.1	59.0
TAA GTA ACC GTT TTC GTC CCA-F	21,012	21	42.9	59.0

Melting temperature (T_m) under LightCycler conditions was calculated using the Tm utility program (Roche Diagnostics).

PCR was performed with the following reaction mix:

	Volume [µl]	[Final]
LightCycler-DNA Master Hybridization Probes	2	1X
$MgCl_2$ (25 mM)	2.4	4.0 mM
Primers (10 µM each)	1+1	0.5 µM each
FITC-labeled probe (4 µM)	1	0.2 µM
LCRed640-labeled probe (8 µM)	1	0.4 µM
H_2O (PCR grade)	9.6	
Total volume	18	

Both 18 µl master mix and 2 µl DNA (50–100 ng/µl) were added to each capillary. Sealed capillaries were spun (1000 x *g* for 20 s) and placed into the LightCycler instrument.

The following PCR protocol was used for amplification:

- Denaturation at 95°C for 2 min
- Amplification

Parameter	Value		
Cycles	45		
Type	Quantification		
	Segment 1	Segment 2	Segment 3
Target temperature [°C]	95	55	72
Incubation time [s]	0	10	5
Temperature transition rate [°C/s]	20	20	20
Acquisition mode	None	Single	None
Fluorimeter gain	F1=1; F2=15; F3=30		

Cooling was done according to the manufacturer's instructions, with a target temperature of 40°C for 30 s.

The number of HHV-6 genomes in each sample per microgram of cellular DNA was calculated.

Results

The validity of this real-time PCR method using the LightCycler instrument was tested in a series of amplification experiments performed on serial dilutions of the reference (Figs. 1, 2). In the range of 10^1–10^5 molecules, there was an inverse linear correlation between the cycle number at the crossing point and the log of the HHV-6 copy number. Therefore, the amount of viral DNA was calculated using this standard curve.

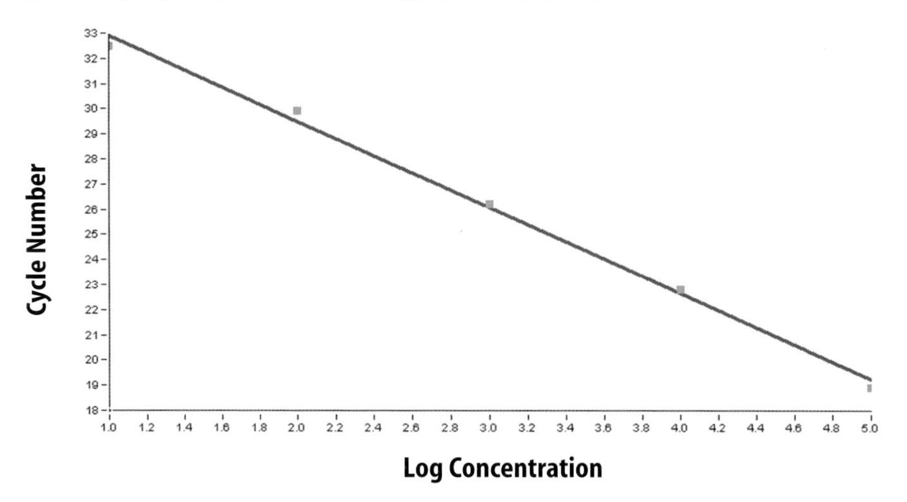

Fig. 1. Standard curve for HHV-6 quantification on the LightCycler. There is a linear correlation in the range of 10^1–10^5 molecules.

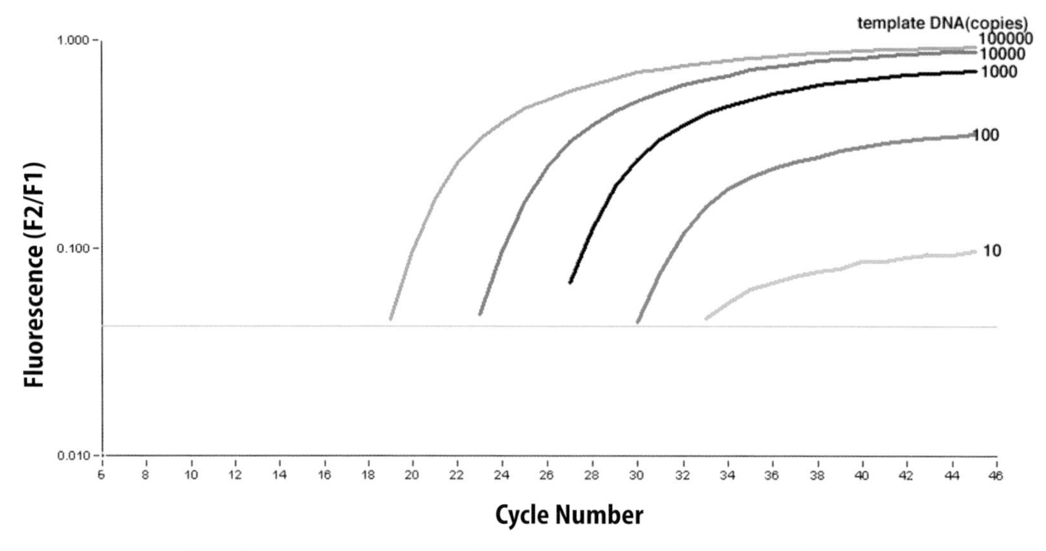

Fig. 2. The fluorescence data used to generate the standard curve in Fig. 1.

In order to determine the HHV-6 viral load in clinical materials detected by real-time PCR, we first tested for viral DNA in peripheral blood samples taken from 12 healthy volunteers. The prevalence of HHV-6 in peripheral blood was 30%; however, the HHV-6 copy number was less than 10^2 copies/assay. Therefore, we considered the normal cut-off level of HHV-6 in the peripheral blood of healthy subjects to be 10^2 copies/assay. We then examined the kinetics of

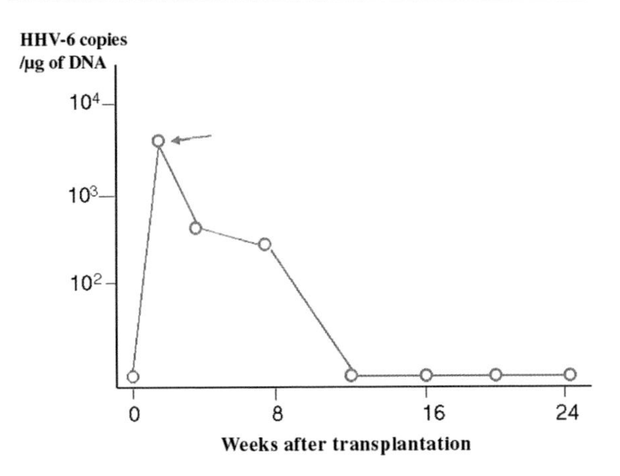

Fig. 3. Kinetics of HHV-6 viral load before and after bone marrow transplantation. PCR on whole blood cells was used for the quantification of HHV-6 genome copies. A remarkable elevation of HHV-6 viral load was found soon after transplantation (*arrowhead*). Delayed bone marrow engraftment was seen in this patient, possibly due to HHV-6 reactivation.

HHV-6 viral load in a patient who underwent allogeneic bone marrow transplantation (Fig. 3). Peripheral blood was collected once before transplantation and once per week after transplantation for 8–24 weeks. We could not detect HHV-6 in the patient before transplantation; however, 2 weeks after transplantation, between 10^3 and 10^4 copies of HHV-6/ug DNA were detected. Levels of HHV-6 copy numbers continued to be high up to 56 days in this patient, and persistent thrombocytopenia was seen. Elevation of HHV-6 antibody titers was not evident in this patient, possibly due to impaired immune status after transplantation. Our study shows that this method may be useful in the management of HHV-6 infection in immunocompromised patients, including transplant recipients.

Comments

To generate a reproducible standard curve, we recommend using plasmid DNA containing the amplified fragment. Alternately, the supernatant of an HHV-6B chronically infected T cell line [Tay(OK)] containing about 10^6 copies/µl can be used as a reference [4]. Another important concern is storage of the hybridization probes. We recommend preparing multiple aliquots in order to avoid repetitive freezing and thawing, and storing the probes at –20°C for less than 6 months.

References

1. Fox JC, Kidd M, Griffiths PD, Sweny P, Emery VC (1995) Longitudinal analysis of cytomegalovirus load in renal transplant recipients using a quantitative polymerase chain reaction: correlation with disease. J Gen Virol 76:309–319
2. Ohyashiki JH, Abe K, Ojima T, Wang P, Zhou CF, Suzuki A, Ohyashiki K, Yamamoto K (1999) Quantification of human herpesvirus 6 in healthy volunteers and patients with lymphoproliferative disorders by PCR-ELISA. Leuk Res 23:625–630
3. Ohyashiki JH, Suzuki A, Aritaki K, Nagate A, Shoji N (2000) Use of real-time PCR to monitor human herpesvirus 6 reactivation after allogeneic bone marrow transplantation. Int J Mol Med 6:427–432
4. Zhou CF, Abe K, Wang P, Ojima T, Yamamoto K (1999) Efficient propagation of human herpesvirus 6B in a T-cell line derived from a patients with adult T cell leukemia/lymphoma. Microbiol Immunol 43:425–436

Simultaneous Identification of Five HCV Genotypes in a Single Reaction

Olfert Landt*, Ulrich Lass, Heinz-Hubert Feucht**

Introduction

Hepatitis C virus (HCV) is considered the major cause of post-transfusion and sporadic non-A, non-B hepatitis with elevated risk for development of liver cirrhosis and hepatocellular carcinoma [1, 2]. The virus is a hepacivirus member of the *Flaviviridae*; HCV is an enveloped single-stranded positive sense RNA virus with a genome of about 9,500 bases. The genome is extremely variable; an infected person harbors billions of virus particles which most likely all carry genomes with different sequences. Eleven genotypes including at least 70 subtypes have been described. Within one subtype the nucleotide similarity is only 76%–80% [3]. Genotypes 1–5 seem to be most important for diagnosis; types 6–11 are very rare and were found only in Asia. Types 1 (80% in Germany) [4] and to a lesser extent 3 and 2 are the most prominent genotypes in Central Europe and North America, type 4 is typical for Northern Africa, type 5 for Southern Africa. Type 3 is often related to i.v. drug abuse [3, 4]. Genotyping of HCV is important because the response of the current therapy regimens and duration of therapy is strongly associated with the genotype [5].

The key for the classification of HCV lies in the choice of the region to be analyzed. The 5′-untranslated region (5-UTR) is the best conserved sequence with only 1%–3% differences within one genotype and 5%–12% between isolates belonging to different types [3]. Therefore, most reverse transcription polymerase chain reaction (RT-PCR)-based diagnostic procedures for virus detection and quantification focus on the well-conserved 5-UTR. To achieve a high sensitivity the amplified products are rather short. Since these fragments are usually available from prior qualitative PCR it makes sense to use them for further analysis [6]. The subtype results from the commercial TRUGENE 5'NC HCV genotyping kit did not correlate with Core DNA sequencing and 5-UTR DEIA data [6]. The 5-UTR contains too little sequence diversity to allow the identification of all known genotypes or even subtyping [3]. We developed a novel genotyping assay based on the 5-UTR which allows the differentiation between genotypes 1–5, but not further subtyping.

* Olfert Landt, Ulrich Lass (✉) (e-mail: Olandt@tib-molbiol.de)
TIB MOLBIOL Berlin, Research and Development Unit, Tempelhofer Weg 11–12, 10829 Berlin, Germany
** Heinz Hubert Feucht
Institut für Medizinische Mikrobiologie, Universitätsklinikum Eppendorf, Martinistraße 52, 20246 Hamburg, Germany

Typing with Hybridization Probes and Multiplexing

Hybridization probes are a sensitive monitor for target sequence deviations covered by the probes. The bound probes are analyzed for their fluorescence decrease during continuous heating; the fluorescence drops when the first oligonucleotide probe detaches from the target. If the pair of probes is made of one tight-binding (anchor) and one less binding probe (sensor), the melting analysis of the probes is solely dependent on the target sequence covered by the sensor.

Two different typing assays can be performed simultaneously in the same capillary, by taking advantage of the dual-channel analysis of the LightCycler in the presence of two dyes. The difference of two melting points for a single base substitution is dependent on the respective base variation between 3 and 10°C. Choosing two different temperature ranges, for example around 50°C and a second above 60°C, allows the combination of two different typing assays in the same reaction and in the same capillary. Running an assay at two temperature levels and two fluorophores enables the analysis of four different sites in one capillary.

Materials

Equipment

LightCycler (Roche Diagnostics, Mannheim, Germany)
LightCycler capillaries (Roche Diagnostics)
Software OLIGO 6.0 for binding energies and folding (Molecular Insights, Cascade, Co., USA)
Software BioEdit 4.0 (Dept. of Microbiology, North Carolina State University [7])
Software mfold at: http://bioinfo.math.rpi.edu
Software BLAST at: http://www4.ncbi.nlm.nih.gov/blast/
Software MOLBIOL or JAVA applet from TIB MOLBIOL for calculation of primer and probe melting temperatures

Reagents

Oligonucleotides and probes (TIB MOLBIOL, Berlin, Germany)
PCR-Buffer (Rapidozym, Berlin, Germany)
Reagent water (MilliQ, Millipore, Eschborn, Germany)

Procedure

Detection Probes for HCV Typing

Public sources of valid genotype-specific sequences for the 5′-untranslated region (5-UTR) are limited. We have aligned sequences that are deposited in GenBank and used these together with alignments kindly provided by Dr. Varnier, Institute of Microbiology in Genoa, Italy, and alignments from Dr. Feucht, Institute for Medical Microbiology in Hamburg, Germany, and the alignments published by Maertens and Stuyer [3].

Base 105 (numbering according to GenBank sequence D10934) is cytosine for genotypes 1, 2, and 5 but uracil for HCV-4. Base 107 is uracil for types 1, 2, and 4 but adenosine for type 5. Guanosine 108 is unique for HCV-1. This permits the selection of a sensor probe specific for HCV-1 and HCV-3 displaying lowered melting temperatures for HCV-2 (one mismatch) or HCV-4 and HCV-5 (each two mismatches). With a second shortened sensor probe specific for genotype 4, we have three mismatches for HCV-5 to achieve a clear differentiation between these two genotypes. HCV-2 has only one mismatch with this probe and genotypes 1 and 3 display two mispaired bases. The anchor probe is identical for all genotypes with exception of cytosine 128 for HCV-3. Although the calculated temperature difference between anchor probe and Sensor 13 was about 10°C, we could detect the base change for HCV-3 working with the synthetic targets containing an adenosine complementary to position 128. In this case the "anchor" probe works as sensor. However, for HCV-3 there are not many 5-UTR sequences published and with clinical samples we observed the same temperature as with HCV-1, indicating that cytosine 128 is not substituted.

The following hybridization probe sets contain two 3′-fluorescein-labeled sensor probes sharing one 5′-LCRed640-labeled probe allowing the identification of four different HCV genotypes:

LC640			D10934	T_m (°C)
Sensor 13	**GTGTCGTGCAGCCTCCAGG-F**	**S**	**100–118**	**62.4**
Sensor 1	GTGTCGTGCAGCCTCCAGG-F	S	100–118	62.4
Sensor 2	GTGTCGTACAGCCTCCAGG-F	S	100–118	56.5
Sensor 3	GTGTCGTGCAGCCTCCAGG-F	S	100–118	62.4
Sensor 4	**TGTTGTACAGCCTCCAGG-F**	**S**	**101–118**	**52.8**
Sensor 5	GTGTCGAACAGCCTCCAGG-F	S	100–118	58.7
Anchor 124	**LCRed640-CCCCCCCTCCCGGGAGAGCC**	**S**	**120–139**	**72.7**
Anchor 3	LCRed640-CCCCCCCTTCCGGGAGAGCC	S	120–139	70.9

Sequences which were used experimentally are printed in bold; the other sequences are shown for comparison only; Base showing variations between the different genotypes are underlined. Melting temperatures (T_m) are calculated with the software MOLBIOL; T_m values are for the shown sequence and not for the mismatch-containing hybrid. Sequence positions are according to GenBank sequence D10934.

Detection Probes for HCV-3a Versus HCV-1

The first set of probes will not provide a clear distinction between genotypes 3 and 4 in the case of HCV-3 samples containing a uracil base at position 128. For HCV-3 samples with 128 cytosine, genotypes 1 and 3 cannot be distinguished. HCV-3a has an adenosine-to-guanosine substitution at position 70 and cytosine instead of uracil at base 78. (With the exception of one single sequence entry for subtype 3g there are only 5-UTR sequences from HCV-3a deposited in GenBank.) In order to overcome these difficulties, we have designed a very short sensor probe covering both exchanged bases with which we detect a fluorescence signal only and exclusively with HCV-3. The anchor probe is labeled with the second dye LCRed705.

LC705			D10934	T_m (°C)
Sensor 1	CACGCAGAAAGCGTCTA-F	S	65–81	50.5
Sensor 2	CACgCAgAAAgCgTCTA-F	S	65–81	50.5
Sensor 3a	**CACGCGGAAAGCGCCTA-F**	**S**	**65–81**	**60.2**
Sensor 3g	CACGCGGAAAGGCCCTA-F	S	65–81	58.9
Sensor 4a	CACGCAGAAAGCGTCTA-F	S	65–81	50.5
Anchor 3	**LCRed705-CCATGGCGTTAGTACGAGTGTC**	**S**	**83–104**	**58.9**

Sequences which were used experimentally are printed in bold; the other sequences are shown for comparison only. Base showing variations between the different genotypes are underlined.Melting temperatures (T_m) are calculated with the software MOLBIOL; T_m values are for the shown sequence and not for the mismatch-containing hybrid. Sequence positions are according to GenBank sequence D10934.

Assay Validation

LightCycler assays can be validated using artificial targets. Quantitation assays are usually tested with cloned PCR fragments. For LightCycler typing assays based on a melting curve analysis of the bound hybridization probes, the target is much shorter and can be synthesized as a long oligonucleotide.

These oligonucleotide targets are reverse complementary to the hybridization probes and elongated by one additional terminal base to avoid side effects from the otherwise resulting blunt ends. With this method we can achieve typical melting data from all target variants described in the literature without having a collection of the respective clinical samples.

In contrast to PCR conditions where hybridization probes are in high excess over the target sequence, we routinely use a threefold excess of target over probe. Although the synthetic target oligonucleotides are HPLC-purified, they are typically 50–80 bases long and a small subset of molecules with wrongly inserted bases, deletions, insertions and oligomers of shorter (*n*-1) sequences can therefore, not be ruled out. These byproducts will yield false lower melting temperature peaks in the analysis. In the presence of excess target molecules, most shorter and false sequences are excluded from probe binding due to their particular lower binding strength.

- Melting Curve Analysis

	Volume [µl]	[Final]
PCR-buffer (10×)	2	1×
$MgCl_2$ stock solution (25 mM)	2.4	3 mM
Target sequence (10 µM)	1.0	0.50 µM
3′-FL hybridization probe (3 µM)	1.0	0.15 µM
5′-640 hybridization probe (3 µM)	1.0	0.15 µM
H_2O (MilliQ, sterile)	12.6	
Total volume	20	

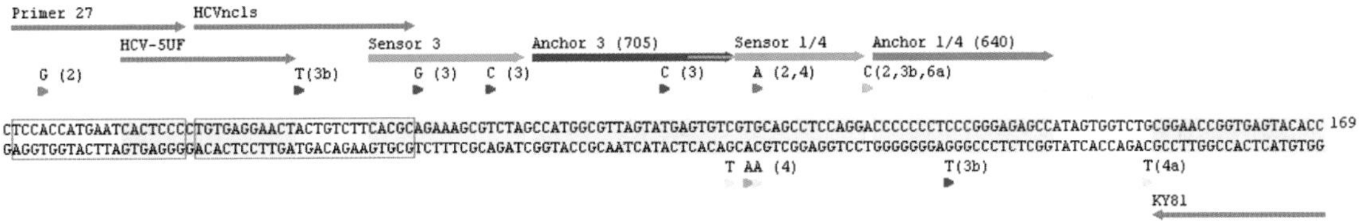

Primer 27
HCVncls
HCV-5UF
Sensor 3
Anchor 3 (705)
Sensor 1/4
Anchor 1/4 (640)
G (2)
T(3b)
G (3)
C (3)
C (3)
A (2,4)
C(2,3b,6a)

CTCCACCATGAATCACTCCCCTGTGAGGAACTACTGTCTTCACGCAGAAAGCGTCTAGCCATGGCGTTAGTATGAGTGTCGTGCAGCCTCCAGGACCCCCCCTCCCGGGAGAGCCATAGTGGTCTGCGGAACCGGTGAGTACACC 169
GAGGTGGTACTTAGTGAGGGGACACTCCTTGATGACAGAAGTGCGTCTTTCGCAGATCGGTACCGCAATCATACTCACAGCACGTCGGAGGTCCTGGGGGGGAGGGCCCTCTCGGTATCACCAGACGCCTTGGCCACTCATGTGG

T AA (4)
T(3b)
T(4a)
KY81

Fig. 1. Amplified sequence with variable bases (triangels). Green arrows: primers for amplification. Orange arrows: detection probes with 3'-fluorescein label. Blue arrows: Anchor with 5'-LightCycler Red label

Melting curve was performed by initial denaturation for 2 min at 95°C, cooling to 45°C and holding for 20 s, slowly heating (0.2 °C/s) from 45°C to 85°C and continuous monitoring of fluorescence.
Typing of PCR products:

	Volume [μl]	[Final]
Buffer (10×)	2.0	1×
$MgCl_2$ stock solution (25 mM)	2.4	3 mM
Primer 27 (10 μM)	1.0	0.50 μM
Primer K81 s (10 μM)	1.0	0.50 μM
Sensor 13 (3 μM)	1.0	0.15 μM
Sensor 4 (3 μM)	1.0	0.15 μM
LC640 Anchor 1 (3 μM)	1.0	0.15 μM
Sensor 3 (3 μM)	1.0	0.15 μM
LC705 Anchor 3 (3 μM)	1.0	0.15 μM
H_2O (MilliQ, sterile)	6.6	
Total volume	18	

Amplification was started with 2 μl of a 1:1,000 diluted diagnostic PCR product made with the identical primers or outer primers. Run conditions were:

Parameter	Value		
Cycles	40		
Type	None		
	Segment 1	Segment 2	Segment 3
Target temperature [°C]	95	52	72
Incubation time [s]	5	12	15
Temperature transition rate [°C/s]	20	20	20
Acquisition mode	None	Single	None
Gains	F1=5; F2=15; F3=30		

Subsequent melting analysis was performed as described previously.

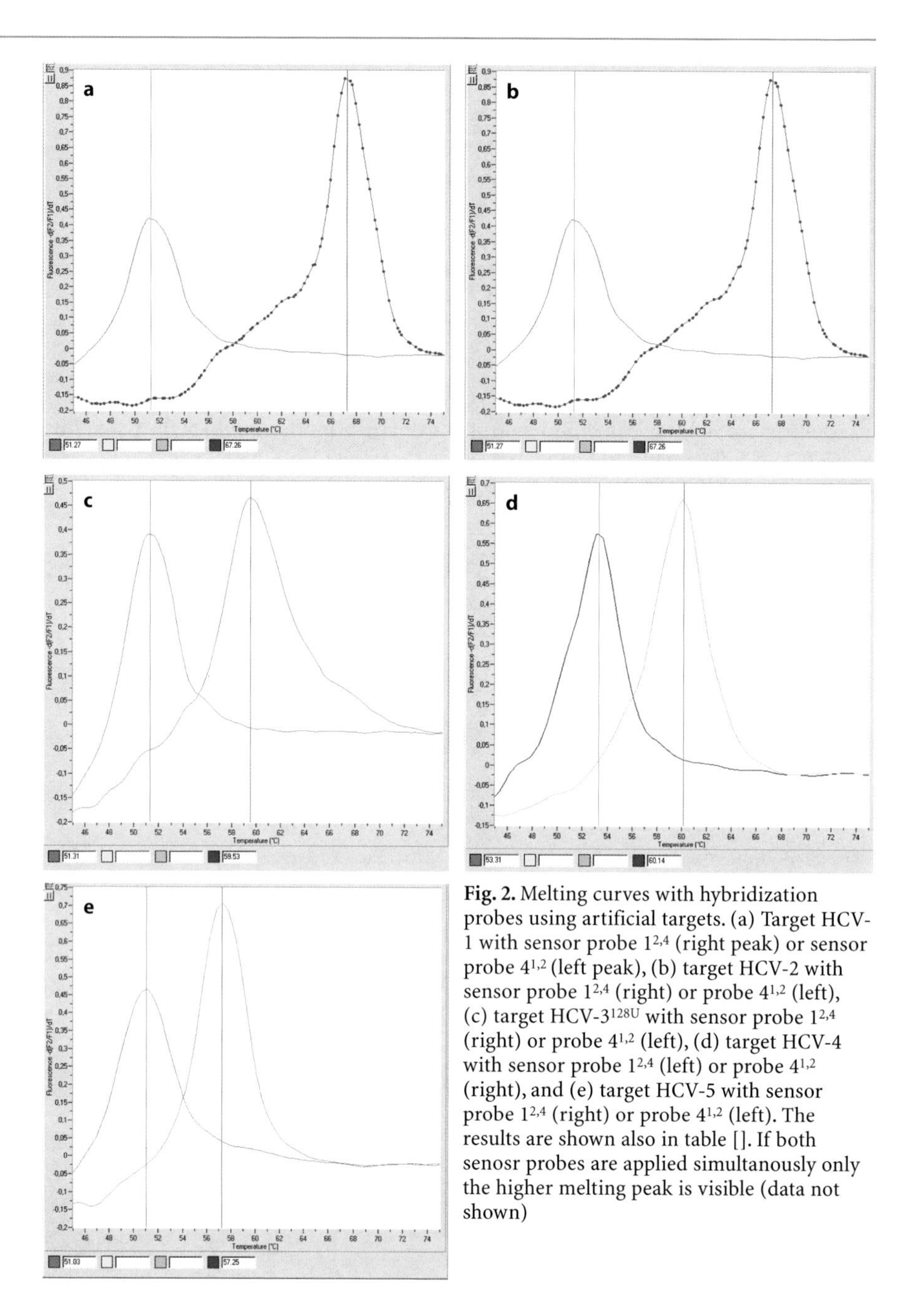

Fig. 2. Melting curves with hybridization probes using artificial targets. (a) Target HCV-1 with sensor probe $1^{2,4}$ (right peak) or sensor probe $4^{1,2}$ (left peak), (b) target HCV-2 with sensor probe $1^{2,4}$ (right) or probe $4^{1,2}$ (left), (c) target HCV-3^{128U} with sensor probe $1^{2,4}$ (right) or probe $4^{1,2}$ (left), (d) target HCV-4 with sensor probe $1^{2,4}$ (left) or probe $4^{1,2}$ (right), and (e) target HCV-5 with sensor probe $1^{2,4}$ (right) or probe $4^{1,2}$ (left). The results are shown also in table [ ]. If both senosr probes are applied simultanously only the higher melting peak is visible (data not shown)

Results

To determine the exact melting temperatures, we analyzed all hybridization probe sets together with each synthetic target sequences separately. In this in vitro experiment we obtained specific melting temperature peaks for every sensor probe and genotype. Due to the competition between the sensor probes, we did not observe the respective lower temperature peak when a mixture of sensor probe 13 and 4 was used. The same was observed in the case of PCR products from pre-typed patient samples.

HCV	Probe	Target sequence	T_m (°C)
1*	13	GCTCTCCCGGGAGGGGGGGTCCTGGAGGCTGCACGACACT	67.4
2	13	GCTCTCCCGGGAGGGGGGGTCCTGGAGGCTGTACGACACT	62.6
3^{128U}	13	GCTCTCCCGGAAGGGGGGG**G**CCTGGAGGCTGCACGACACT	59.4
4	13	GCTCTCCCGGGAGGGGGGGTCCTGGAGGCTGTACAACACT	57.3
5	13	GCTCTCCCGGGAGGGGGGGTCCTGGAGGCTGTTCGACACT	53.2
1	4	GCTCTCCCGGGAGGGGGGGTCCTGGAGGCTGCACGACACT	51.0
2	4	GCTCTCCCGGGAGGGGGGGTCCTGGAGGCTGTACGACACT	57.8
3^{128U}	4	GCTCTCCCGGAAGGGGGGG**G**CCTGGAGGCTGCACGACACT	51.1
4	4	GCTCTCCCGGGAGGGGGGGTCCTGGAGGCTGTACAACACT	60.3
5	4	GCTCTCCCGGGAGGGGGGGTCCTGGAGGCTGTTCGACACT	50.6

* Sequence is identical to HCV-3 containing 128 cytosine. The second base difference at position 119 (bold) is located in the gap between both hybridization probes and does not interfere with the assay. Variable bases are underlined.

Measured melting temperatures (T_m in °C in the PCR-free system):

HCV	Sensor 13	Sensor 4	Sensor 3a (705)
1	**67.3**	51.3	–
2	**62.5**	57.6	–
3^{128C}	**67.3***	51.3	**60**
3^{128U}	**59.5**	51.3	**60**
4	53.3	**60.1**	–
5	**57.2**	51.0	

In the above table, results from melting curves using artificial targets are shown. The highest temperature observed for each dye is shown in bold letters; the lower temperature is not expected to appear in the assay containing both sensor probes 13 and 4. HCV-3^{128C} cannot be distinguished from HCV-1, and HCV-3^{128U} shows very similar melting temperatures to HCV-4. The T_m value for HCV-2^{128C} is taken from the HCV-1 measurement.

Observed melting temperatures (T_m in °C from HCV PCR products):

HCV^{PCR}	LC640	LC705
1	66.5	–
2	63.0	–
3	66.0	61
4	60.7	–

Melting temperatures observed from PCR product targets using a mixture of hybridization probes 13 and 4 in channel F2 (LC640) and probe 3a in channel F3 (LC705). An HCV-5 sample was not available for testing.

With the described hybridization probes, all five HCV genotypes can be distinguished by their specific melting temperature (Table 4).

Table 5. Oligonucleotides

	Primers and Probes for the amplification of the 5-UTR		D10934	T_m (°C)
Primer 27	TCCACCATgAATCACTCCC	S	26–44	54.1
HCVncl s	TGTGAGGAACTWCTGTCTTCACGC	S	46–69	60.9
KY78s	CAAGCACCCTATCAGGCAGT	A	308–288	57.2
KY81 s	GGTGTACTCACCGGTTCCG	A	169–151	58.4
	Probes and controls for genotypes 1/2/4/5			
Sensor 1[2,4]	GTGTCGTGCAGCCTCCAGG-F	S	100–118	62.4
Sensor 4[1,2]	TGTTGTACAGCCTCCAGG-F	S	101–118	52.8
Anchor 1	LCRed640-CCCCCCCTCCCGGGAGAGCC	S	120–139	72.7
HCV 1	GCTCTCCCGGGAGGGGGGGTCCTGGAGG CTGCACGACACT	A	140–99	
HCV 2	GCTCTCCCGGGAGGGGGGGTCCTGGAGG CTGTACGACACT	A	140–99	
HCV 3	GCTCTCCCGGAAGGGGGGGGCCTGGAGG CTGCACGACACT	A	140–99	
HCV 4	GCTCTCCCGGGAGGGGGGGTCCTGGAGG CTGTACAACACT	A	140–99	
HCV 5	GCTCTCCCGGGAGGGGGGGTCCTGGAGG CTGTTCGACACT	A	140–99	
	Probes and controls for genotype 3a			
Sensor 3	CACGCGGAAAGCGCCT-F	S	65–81	60.2
Anchor 3	LCRed705-CCATGGCGTTAGTACGAGTGTC	S	83–104	58.9
HCV 1	CGACACTCGTACTAACGCCATGGCTAGACGCTTT CTGCGTGAA105–64HCV 3CGACACTCATACTAA CGCCATGGCTAGGCGCTTTCCGCGTGA	A	105–64	

Comments

This assay includes base variations which are partly covered by the commonly used forward primer KY80, and the respective PCR products including one from the Roche AMPLICOR kit cannot be used for this analysis. We recommend using outer forward primers such as "primer 27" or "HCVn1" and the primer KY78 (or the shorter variant KY78s, product is a 263- or 283-bp fragment) or the inner primer KY81s (144- or 124-bp fragment) for the RT-PCR amplification. A comparative study between 5UTR sequencing results and the described LightCycler assay is in press.

References

1. Schreiber GB, Busch MP, Kleinmann SH, Korelitz JJ (1996) The risk of transfusion-transmitted viral infections. N Engl J Med 334:1695–1690
2. Takada A, Tsutsumi M, Zhang SC, Okanoue T, Matsushima T, Fujiyama S, Komatsu M (1996) Relationship between hepatocellular carcinoma and subtypes of hepatitis C virus: a nationwide analysis. J Gastroenterol Hepatol 11:166–169
3. Maertens G, Stuyer L (1997)Genotypes and genetic variation of hepatitis C virus. In: Harrison TJ Zuckermanm AJ (eds) The molecular medicine of viral hepatitis. John Wiley, Chichester, New York, Weinheim, Brisbane, Toronto, Singapore 183–234
4. Ross RS, Viazov S, Renzing-Kohler K, Roggendorf M (2000) Changes in the epidemiology of hepatitis C infection in Germany: shift in the predominance of hepatitis C subtypes. J Med Virol 60:122–125
5. Zein, NN, J Rakela, EL Krawitt, KR Reddy, T Tominaga, DH Persing, and the Collaborative Group (1996) Hepatitis C virus genotypes in the United States: epidemiology, pathogenicity and response to interferon therapy. Ann Intern Med 25:634–639
6. Ross RS et al (2000) Genotyping of Hepatitis C virus isolates using CLIP sequencing. J Clin Microbiol 38:3581–3584
7. Hall TA (1999) BioEdit: a user-friendly biological sequence alignment editor and analysis program for Windows 95/98/NT. Nucl Acids Symp Ser 41:95–98

Rapid Detection of West Nile Virus

Olfert Landt*, Jasmin Dehnhardt**, Andreas Nitsche*,
Gary Milburn***, Shawn D. Carver***

Introduction

West Nile virus (WNV) is a flavivirus endemic in Africa, the Middle East, and in South-western Asia. The virus is a member of the Japanese encephalitis serocomplex, containing a plus-strand ssRNA genome of about 11,000 bases. Natural hosts of the virus are reported to be mammals and also birds, and infections can be transmitted through mosquitoes [1]. Human infections are usually mild or subclinical, but during two recent epidemics in Russia and North America several mortal cases of encephalitis occurred [2]. Serological tests do not differentiate between WNV and related members of the same serocomplex, and virus isolation in cell culture followed by immunofluorescence-based identification takes about 1 week. Diagnostic assays based on reverse transcription polymerase chain reaction (RT-PCR) of the virus genome allow a quicker analysis [3–5]. Immediate pathogen identification in cases of encephalitis of unknown origin can help to prevent future epidemics.

A few RT-PCR based diagnostic assays have been previously described [3]. Briese et al. reported two TaqMan probe-based real-time PCR assays using the genomic region for the nonstructural proteins NS3 and NS5 [4]. Since several WNV genomic sequences are deposited in GenBank, we aligned the published NS primers with the GenBank sequences and constructed new consensus primers and hybridization probes for the described regions (Table 1). However, the amplification efficiency and detection limit for our improved NS3 and NS5 PCR assays were not satisfactory for diagnostic purposes. Lanciotti et al. used primers optimized for the New York 1999 isolates for a TaqMan probe-based assay in the 3′-noncoding (3NC) region and a second assay in the envelope protein gene [5]. We have developed a new hybridization probe-based assay targeting the published 3NC sequences.

* Olfert Landt, Andreas Nitsche (✉) (e-mail: Olandt@tib-molbiol.de)
TIB MOLBIOL Berlin, Research and Development Unit, Tempelhofer Weg 11–12, 10829 Berlin, Germany

** Jasmin Dehnhardt, Gen Express GmbH, Tempelhofer Weg 11–12, 10829 Berlin, Germany

*** Gary Milburn, Shawn D. Carver, American Medical Laboratories, Molecular Biology Department, 14225 New Brook Drive, Chantilly, VA 20151-2230, USA

Table 1. Oligonucleotides

TaqMan[a]	(103 bp)		AF202541	T_m (°C)
WN39NC F	CAGACCACGCTACGGCG	S	10,626–10,642	59.7
WN39NC R	CTAGGGCCGCGTGGG	A	10,728–10,714	59.3
WN39NC TM	6FAM-TCTGCGGAGAGTGCAG TCTGCGA(XT)P	S	10,649–10,672	67.8
LightCycler	**(98 or 110 bp)**		**AF202541**	**T_m (°C)**
WN39NC S	CCCAATGTCAGACCACGCT	S	10,619–10,636	58.7
WN39NC B	GGGGCGTTGGTTTGCCT	A	10,716–10,700	61.6
WN39NC R[a]	CTAGGGCCGCGTGGG	A	10,728–10,714	59.3
WN39 FLU	CTGGGGCACTATCGCAGACTGC-FL	A	10,682–10,661	64.9
WN39 LCR	LC640-CTCTCCGCAGAGTAG CACGCCG P	A	10,659–10,638	67.7

[a] Lanciotti et al. [5].
FL, fluorescein; p, phosphate; LC640, LightCycler Red 640, (XT), TAMRA-dThymidin.
Melting temperatures calculated with the program MOLBIOL.

Materials

Equipment

LightCycler (Roche Diagnostics, Mannheim, Germany)
LightCycler capillaries (Roche Diagnostics)
Eppendorf MasterCycler (Eppendorf, Hamburg, Germany)
Software OLIGO 6.0 for binding energies and folding (Molecular Insights, Cascade, Co., USA)
Software mfold at: http://bioinfo.math.rpi.edu
Software BLAST at: http://www4.ncbi.nlm.nih.gov/blast/
Software MOLBIOL or JAVA applet from TIB MOLBIOL for calculation of primer and probe melting temperatures (Berlin, Germany)

Reagents

LightCycler – RNA Amplification Kit Hybridization Probes (Roche Diagnostics, Branchburg, NJ, USA)
MilliQ Water (Millipore, Eschborn, Germany)
QIAGEN Viral RNA MINI KIT (Qiagen, Valencia, CA, USA)
NucleoSpin plasmid kit (Machery-Nagel, Rapidozym, Berlin, Germany)
ATCC VR-82 (American Type Culture Collection, Manassas, VA, USA)
Platinum *Taq* Polymerase (Life Technologies, Eggenstein, Germany)
GeneTherm *Taq* Polymerase (Diagnostic Research, Athens, Greece)
dNTPs (Amersham Pharmacia, Freiburg, Germany)
DMSO (Amersham Pharmacia)
BSA (Invitrogen, Karlsruhe, Germany)
Oligonucleotides and probes (TIB MOLBIOL, Berlin, Germany)
Cloning services (GenExpress, Berlin, Germany)

TA cloning vector (Invitrogen)
BigDye Terminator Cycle Sequencing Kit (Applera Deutschland, Weiterstadt, Germany)

Procedure

Primer and Hybridization Probe Design

The efficiency and the detection limit of a given PCR system are mainly dependent on the selection of the target sequence and the amplification primers, and less dependent on the sequence of the detection probes. Virus genomes can be highly variable and the development of diagnostic PCR assays is limited to well-conserved regions. A total of 194 West Nile Virus (WNV) sequences are reported in GenBank. The region for the WNV assay described here was selected according to the published assay from Lanciotti. The published primers were analyzed for their target specificity; particularly the 3′-termini were examined for their potential to initiate primer dimers and shorter products. The GenBank AF202541 WNV sequence was analyzed for secondary structures in the target region (Fig. 1). This is of greater importance for RNA targets due to their more rigid structures. We found a potential stem loop structure (CACGC-GCGTG) with a calculated melting temperature (T_m) of 72°C (mfold) or even 89°C (OLIGO 6.0) covering the binding site for the forward primer as published by Lanciotti. Furthermore, the termini of the published primers contain up to five consecutive GC bases. It is our experience that primers ending on A or T bases and/or containing less G/C bases at their 3′-termini hybridize more accurately, leading to a higher sensitivity for low-abundant targets.

The forward and reverse primers we have selected are both shifted a few bases upstream. A GenBank search confirmed their specificity for WNV. We found eight WNV sequence matches to our forward primer (primer S) and six matches to our reverse primer (primer B), contrasting the six matches and six matches and two sequence deviations for the respective previously published primers. Primers S and B give rise to a 98-bp amplicon, five bases shorter than the product reported by Lanciotti et al. [5]. The resulting amplicon spanned by the primer pair S and B is five bases shorter than the 103-bp product used by Lanciotti.

The hybridization probes were placed at the 3′-end of the lower strand, having a maximum distance from the primer on the same strand. The 3′-fluorescein-labeled probe has 64% GC and 50% purine (A, G) bases, the 5′-LCRed-labeled probe has 65% GC and less than 50% purine bases.

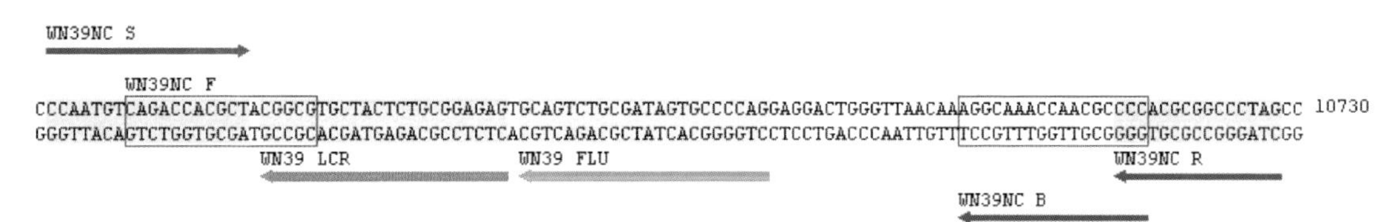

Fig. 1. Overview on the location of primers and hybridization probes (sequence numbering according to AF202541)

Cloning of a Synthetic Standard

For the development of this assay we prepared a synthetic DNA fragment, which was constructed in a PCR fusion reaction from long oligonucleotide targets, using each 0.5-pmol WEST [5′-CCTCCTggggCACTATCgCAgACTgCACTCTCCgCAgAg-TAgCACgCCgTAg] and EAST [5′-gCCCCAggAggACTgggTTAACAAAggCAAAC-CAACgCCCCACgCg] and each 20 pmol of the amplification primers WN3NC S+ [5′-CCCAATgTCAgACCACgCTACggCgT] and WN3NC R [5′-CTAgggCCgCgTggg] in a standard 50 μl reaction, running 30 cycles with 95°C denaturation, 10 s annealing at 58°C, and 10 s extension reaction at 72°C. The resulting product was cloned in a TA-cloning vector following the manufacturer's instructions. Positive clones were expanded, plasmid DNA was isolated using a NucleoSpin plasmid kit, and the sequence was confirmed on an Applied Biosystems 310 capillary sequencer.

Preparation of Template RNA

Virus RNA was isolated from 140-μl patient samples using the Qiagen Viral RNA kit following the manufacturer's instructions. The final elution volume was 60 μl.

Plasmid Standard and Virus Reference

The plasmid sample was quantified spectrophotometrically and dilutions containing 10–10,000,000 DNA genome equivalents were prepared (Fig. 2). As virus sample we used the VR-82 virus standard from the American Type Culture Collection. This strain was isolated in 1937 in Uganda. For the preparation of viral RNA we used 5 μl of the ATCC standard diluted in 145 μl water. The RNA was extracted as described above and used to prepare a dilution row.

Inhibition Control

For routine use, we run a virus-spiked patient sample as inhibition control.

LightCycler PCR

The PCR was established using the cloned plasmid controls. The assay was compared with the published TaqMan assay using virus-spiked patient probes.

Run conditions (plasmid control):

	Volume [μl]	[Final]
Native *Taq* Polymerase 5 U/μl	0.5	0.1 U/μl
PCR buffer as supplied (10×)	2.0	1×
dNTPs (10 mM)	1.0	0.50 μM
BSA 4 μg/μl	1.0	0.20 μg/μl
$MgCl_2$ stock solution (25 mM)	3.2	4.00 mM
Primer S (10 μM)	0.5	0.25 μM
Primer B (10 μM)	0.5	0.25 μM
3′-FL hybridization probe (3 μM)	1.0	0.15 μM
5′-640 hybridization probe (3 μM)	1.0	0.30 μM
Optional use of 3% DMSO	–	2%–5%
H_2O (MilliQ, sterile)	6.3	
Total volume	18	

In total, 18 μl of the master mix and 2 μl of the plasmid dilution were added to each capillary. The capillaries were sealed, spun down, and placed into the LightCycler

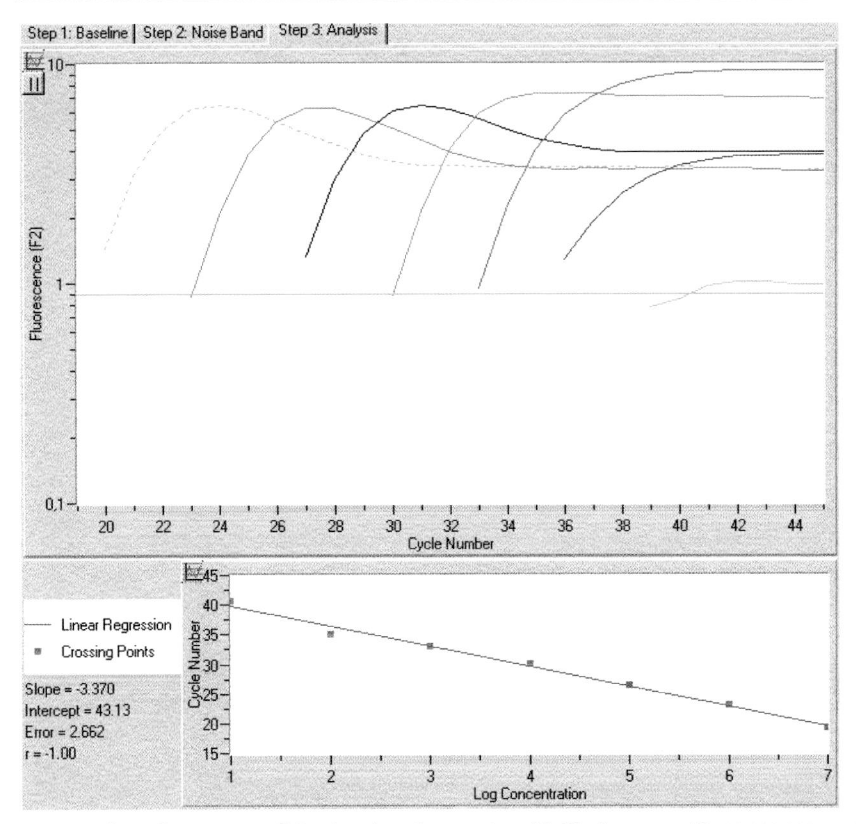

Fig. 2. LightCycler DNA amplification data from a plasmid dilution row with 10,000,000 copies (*green dotted*) to 100 copies (*deep blue*) and 50 copies (*light green*). Original fluorescence values are between 22 and 30 units (gain 15)

rotor. The following PCR protocol was used for amplification and hybridization probe-based detection of the plasmid dilutions:

- Initial denaturation for 3 min at 95°C
- Amplification

Parameter	Value		
Cycles	45		
Type	None		
	Segment 1	Segment 2	Segment 3
Target temperature [°C]	95	56	72
Incubation time [s]	6	10	14
Temperature transition rate [°C/s]	20	20	20
Acquisition mode	None	Single	None
Gains	F1=1; F2=15; F3=30		

- Cooling for 30 s at 40°C

Run conditions for the RT-PCR TaqMan assay in the LightCycler instrument:

	Volume [µl]	[Final]
LC – RNA Amplification Kit Mix	2.0	1×
$MgCl_2$ stock solution (25 mM)	4.8	6+1.5* mM
Primer S (10 µM)	1.0	0.50 µM
Primer B (10 µM)	1.0	0.50 µM
TaqMan probe (5 µM)	1.0	0.25 µM
H_2O (MilliQ, sterile)	8.2	
Total volume	18	

* 1.5 mM $MgCl_2$ contained in the amplification mix

In total, 18 µl of the master mix and 2 µl of the corresponding template preparation were added to each capillary. The capillaries were sealed, spun down, and placed into the LightCycler rotor. The following PCR protocol was used for amplification and TaqMan detection:

- Reverse transcription at 55°C for 30 min
- Denaturation for 2 min at 95°C
- Amplification

Parameter	Value		
Cycles	45		
Type	None		
	Segment 1	Segment 2	Segment 3
Target temperature [°C]	95	57	68
Incubation time [s]	5	15	30
Temperature transition rate [°C/s]	20	20	20
Acquisition mode	None	None	Single
Gains	F1=5; F2=15; F3=30		

- Cooling for 30 s at 40°C

Run conditions for the RT-PCR assay with hybridization probes (clinical specimen):

	Volume [µl]	[Final]
LC RNA amplification mix	2.0	1×
$MgCl_2$ stock solution (25 mM)	4.8	6+1.5* mM
Primer S (10 µM)	1.0	0.50 µM
Primer B (10 µM)	1.0	0.50 µM
3′-FL hybridization probe (4 µM)	1.0	0.20 µM
5′-640 hybridization probe (4 µM)	1.0	0.20 µM
H_2O (MilliQ, sterile)	7.2	
Total volume	18	

*1.5 mM $MgCl_2$ contained in the amplification mix

In total, 18 μl of the master mix and 2 μl of the corresponding template preparation were added to each capillary. The capillaries were sealed, spun down, and placed into the LightCycler rotor. The following PCR protocol was used for amplification and hybridization probe-based detection:

- Reverse transcription at 55°C for 30 min
- Denaturation for 30 s at 95°C
- Amplification

Parameter	Value		
Cycles	45		
Type	None		
	Segment 1	Segment 2	Segment 3
Target temperature [°C]	95	57	72
Incubation time [s]	5	15	25
Temperature transition rate [°C/s]	20	20	20
Acquisition mode	None	Single	None
Gains	F1=5; F2=30; F3=30		

- Cooling for 30 s at 40°C

Results

Using the plasmid dilutions we observed a rather strong Hook effect, especially for higher-concentrated samples, a phenomenon typical for short amplicons with high GC content. The fast re-annealing of the PCR strands competes with probe binding, thus lowering the average fluorescence. We have used twofold higher concentrated acceptor probes to overcome this effect and to achieve higher fluorescence values. Alternatively, an asymmetric PCR could be used, but changing primer concentrations often lowers the absolute sensitivity. However, the Hook effect does not interfere with the crossing points and therefore does not change the quantification results. The linear detection range was found to be 50–10,000,000 genome equivalents. Addition of 2%–5% DMSO did not lower the Hook effect significantly but enhanced the sensitivity to ten genome equivalents per reaction. With virus-spiked patient samples we achieved an at least tenfold higher sensitivity compared with the published TaqMan assay on the LightCycler instrument. Most of the other assays we have studied so far, for example the detection of the human cytomegalovirus with TaqMan and hybridization probes [6], displayed identical sensitivities. We, therefore, believe that the observed enhanced sensitivity is mainly due to our selection of improved primer sequences. We have not yet used the altered primers together with the published TaqMan probe.

Comments

Previous real-time assays for the detection of West Nile virus were based on the use of TaqMan probes. In this publication we report comparable if not higher sensitivities achieved by using hybridization probes. From the point of sequence conservation we prefer to use hybridization probes because of their lower sensitivity towards single base changes. In the case of virus genome variations, the melting temperature of the hybridization probes can be lowered below the annealing temperature, thus losing the potential to measure real-time quantification; but in contrast to a TaqMan-based assay, one can still detect a specific signal running an additional melting curve analysis.

We did not analyze the specificity of the described primers for different strains or isolates experimentally. Other primer combinations than those described here could enhance the detection of a greater variety of WNV isolates.

References

1. Anderson JF, Andreadis TG, Vossbrinck CR, Tirrell S, Wakem EM, French RA, Garmendia AE, Van Kruiningen HJ (1999) Isolation of West Nile virus from mosquitoes, crows, and a Cooper's hawk in Connecticut. Science 286:2331–2333
2. Asnis D, Conetta R, Waldmon G, et al (1999) Outbreak of West Nile-like viral encephalitis – New York. Centers for disease control and prevention. Morb Mortal Wkly Rep 48:845–849
3. Porter KR, Summers PL, Dubois D, Puri B, Nelson W, Henchal E, Oprandy JJ, Hayes CG (1993) Detection of West Nile virus by the polymerase chain reaction and analysis of nucleotide sequence variation. Am J Trop Med Hyg 48:440–446
4. Briese T, Glass WG, Lipkin WI (2000) Detection of West Nile virus sequences in cerebrospinal fluid. Lancet 355:1614–1615
5. Lanciotti et al. (2000) Rapid detection of West Nile virus from human clinical specimens, field-collected mosquitoes, and avian samples by a TaqMan reverse transcriptase-PCR Assay. J Clin Microbiol 38:4066–4071
6. Nitsche A, Steuer N, Schmidt CA, Landt O, Siegert W (1999) Different real-time PCR formats compared for the quantitative detection of human cytomegalovirus DNA. Clin Chem 45:1932–1937

Use of Rapid Cycle Real-Time PCR for the Detection of Rabies Virus

LORRAINE MCELHINNEY*, JASON SAWYER,
CHRISTOPHER J. FINNEGAN, JEMMA SMITH, ANTHONY R. FOOKS

Introduction

Classical rabies virus (RV) is a member of the *Lyssavirus* genus within the *Rhabdoviridae* family, and is a causative agent of rabies. The virus has a wide host range, most probably including all mammals.

The genome is composed of a non-segmented, negative sense RNA strand approximately 12,000 nucleotides in length. This encodes five separate proteins designated (N) nucleoprotein, (M1) phosphoprotein, (M2) matrix protein, (G) glycoprotein and (L) polymerase [1]. The amino acid sequences of the N protein display a high degree of homology [2] and hence, the RV N-gene acts as an appropriate target sequence for use in nucleic acid detection systems for rabies.

We have adapted our existing rabies virus and 18S rRNA RT-PCR detection systems for use on the LightCycler instrument (Roche Diagnostics, UK). Amplification of 18S rRNA, a control for the integrity of RNA, generates a clear signal indicating that the RT-PCR has been successful [3]. Currently, the development of a system for quantitative measurement of RV in clinical samples using LightCycler assays is underway in our laboratory.

This assay provides rapid analysis without the need for subsequent gel electrophoresis. The system also provides the ability to develop a quantitative assay for viral detection based on threshold detection of the amplified product. The use of the commercial Ambion Competimer system allows the amplification signal of the abundant 18S rRNA target to be attenuated to levels typically observed when detecting RV by PCR. We have developed a RV nucleoprotein RNA transcription vector to provide known amounts of RNA target for use as quantification controls.

Materials

Equipment

LightCycler instrument (Roche Diagnostics, UK)

Reagents

TRIzol (Gibco-BRL, The Netherlands)
Amplification primers (MWG-Biotech, Germany)

* Lorraine McElhinney (✉) (e-mail: L.McElhinney@vla.defra.gsi.gov.uk)
Rabies Research and Diagnostics Group, Veterinary Laboratories Agency (Weybridge),
New Haw, Addlestone, Surrey KT15 3NB, UK

Ambion QuantumRNA 18S Internal Standards (AMS Biotechnology, UK)
TaqMan probes (Oswel, Southampton, UK)
LightCycler – RNA Amplification Kit Hybridization Probes (Roche Diagnostics)
HPLC grade water (Sigma, UK)

Procedure

RNA Propagation and Extraction RNA was extracted directly from all infected and uninfected brain material using TRIzol according to the manufacturer's instructions (Gibco-BRL).

Primers Specific RV primers were designed to anneal to the nucleoprotein gene of classical RV as described [4]. Primers from the Ambion QuantumRNA 18S Internal Standards kit (Ambion) were used for amplification of internal control 18S rRNA molecules. TaqMan probes were designed for recognizing the RV nucleoprotein 18S rRNA amplicons. PCR primers specific for both RV nucleoprotein and murine 18S rRNA are shown in Table 1.

LightCycler RT-PCR Rabies virus RT-PCR was performed with the following reaction mix:

	Volume [µl]	[Final]
LightCycler RT-PCR reaction mix hybridization probes	2	1×
$MgCl_2$ (25 mM)	2.4	6 mM
RT-PCR enzyme mix	0.4	1×
Primer 1 (7.5 µM)	1.6	0.6 µM
Primer 2 (7.5 µM)	1.6	0.6 µM
TaqMan probe (5 µM)	0.6	0.15 µM
H_2O (HPLC grade)	10.4	
Total volume	19	

18S rRNA RT-PCR was performed with the following reaction mix:

	Volume [µl]	[Final]
LightCycler RT-PCR reaction mix hybridization probes	2	1×
$MgCl_2$ (25 mM)	2.4	6 mM
RT-PCR enzyme mix	0.4	1×
Primers (5 µM each)	2.0	0.5 µM
TaqMan probe (5 µM)	0.6	0.15 µM
H_2O (HPLC grade)	11.6	
Total volume	19	

We added 19 µl of master mix and 1 µl of RNA (20–50 ng) to each capillary. Sealed capillaries were spun (1000 *g* for 15 s) and placed in the LightCycler instrument.

The following RT-PCR protocol was used for amplification:

- Reverse transcription at 50°C for 10 min
- Denaturation at 95°C
- Amplification

Parameter	Value	
Cycles	40	
Type	Quantification	
	Segment 1	Segment 2
Target temperature [°C]	95	58
Incubation time [s]	1	30
Temperature transition rate [°C/s]	20	20
Acquisition mode	None	Single
Gains	F1=1; F2=15	

Table 1. Oligonucleotides

Rabies nucleoprotein (GenBank Accession #X03673)				
	Position	Length	GC (%)	T_m (°C)
PCR primers				
ATG TAA CAC CYC TAC AAT G	55–73	19	37	57.8
CAA TTC GCA CAC ATT TTG TG	660R–641	20	40	59.3
PCR product	55–660	606		
TaqMan probe				
F-CCC AAT TCC CGT CTA CAT CAG TAC GC-T	360R–335	27	52	70.8
Mouse 18S rRNA (GenBank Accession # X00686)				
PCR primers				
TCA AGA ACG AAA GTC GGA GG	1026–1045	20	50	63.2
GGA CAT CTA AGG GCA TCA CA	1494–1513R	20	50	62.8
PCR product	1026–1513	488		
TaqMan probe				
F-CGT AGT TCC GAC CAT AAA CGA TGC CG-T	1068–1094	27	52	71.5

F, 5′ terminal fluorescein; T, 3′ terminal TAMRA.

Results

Detection of a laboratory standard, challenge virus strain (CVS 11) viral RNA, extracted from infected mouse brain, is illustrated in Fig. 1. Addition of variable amounts of target RNA led to changes in crossing point values, illustrating the potential of using this system for the quantitative measurement of RV.

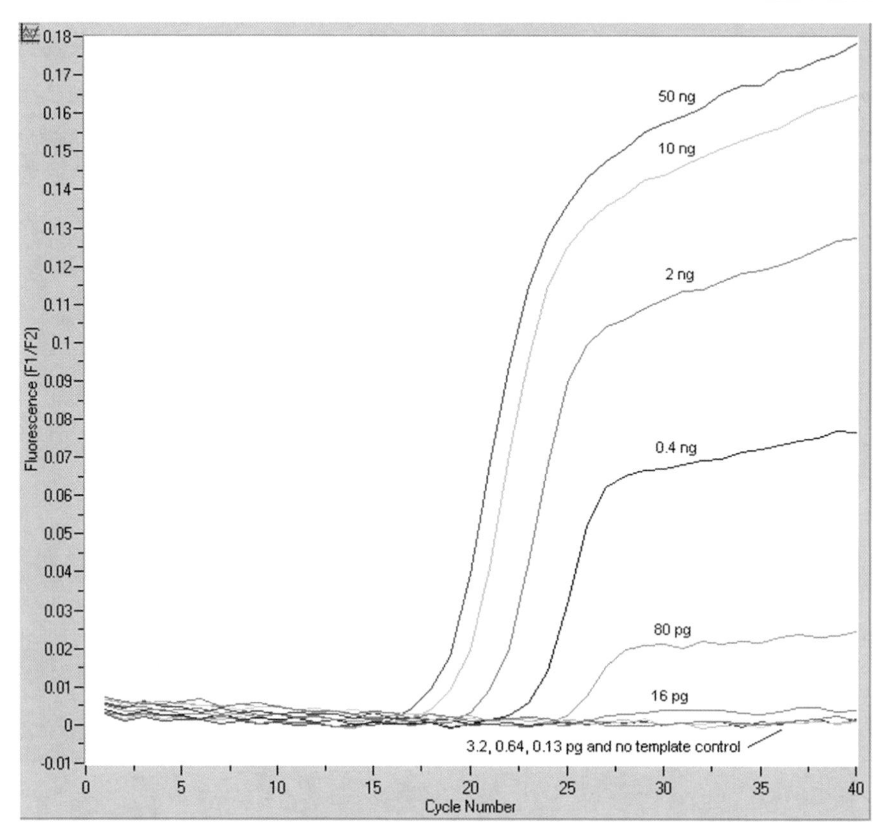

Fig. 1. RT-PCR of a dilution series of RNA extracted from RV-infected mouse brain using RV-specific primers. The figure shows the amount of mouse brain total RNA added to each reaction

In order to achieve accurate quantification, it is necessary to take into account sample variables such as differences in RNA degradation and RNA yield. This can be achieved by concurrent amplification of an internal control. Our previous use of the Ambion QuantumRNA 18S internal standards has therefore been extended to this work. A TaqMan probe was designed to allow real time detection of the 18S rRNA amplicon (Fig. 2).

Furthermore, we have made use of the Competimer System, which forms part of the Ambion QuantumRNA 18S kit. Competimers are modified primers that cannot be extended. By adjusting the ratio of competimers to normal 18S rRNA primers, the amplification signal can be attenuated to a level identical to that of the intended target. In real-time PCR, this allows manipulation of the crossing point value (Fig. 2). In this way, 18S rRNA amplification signals can be attenuated to crossing point values typical of those measured for RV.

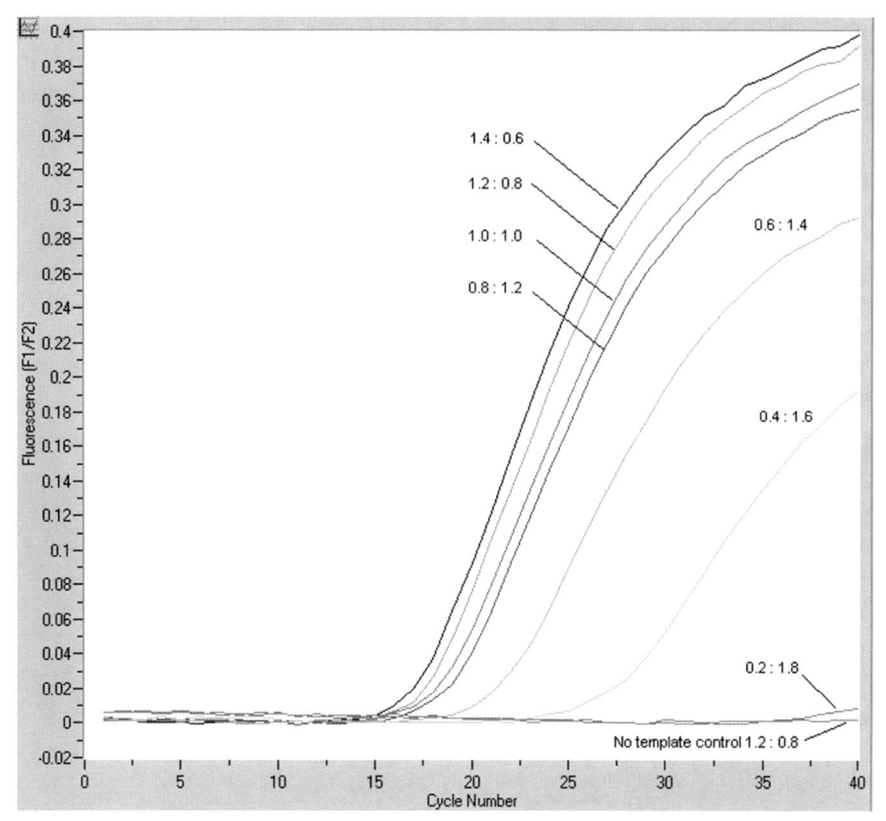

Fig. 2. RT-PCR of RNA extracted from mouse brain using different ratios of Competimers for 18S rRNA detection. The figure shows the ratio of primers:Competimers used in each reaction

Comments

The quantification system we are developing will be used primarily to determine the total rabies viral load in an infection model. Currently, different approaches to quantification of rabies virus are underway in our laboratory, however, it is still unclear which approach will produce the most practical and useful technique.

Use of Competimer System

The first approach is quantification relative to the 18S rRNA gene. At present, we amplify the rabies specific primers and 18S rRNA primers in different tubes. We are currently working towards amplifying both the target and control in a one-tube system employing two different dyes. We have used the Competimer system in this study since the signal we obtained from initial 18S rRNA amplification was much stronger than the viral signal. This effect also occurred for tissue samples where the viral load was expected to be high (cross point values of ~10 compared to ~ 20–25 respectively). Work is underway to balance the relative amplification of the internal control and target in the same tube. However, if we continue to use separate tubes for both target and degradation control, this will not be necessary and it might be that the dynamic range of both control and target primers are broad enough not to need the use of competimers.

Use of Synthetic Viral Transcript

The second approach is to amplify samples against standard curves made up of known amounts of viral target in a background of mouse total RNA. The viral transcript was developed by employing the vector pCR2.1 (Invitrogen, UK.) containing a sequence encoding the RV nucleoprotein. Transcripts were then generated using the appropriate polymerase. We hope to use the RNA transcript system to produce known amounts of viral RNA target that can be used to spike negative RNA samples. In this way we aim to be able to produce artificial samples that mimic natural samples.

Use of RNA Loading Control

Lastly, we have been assessing the use of the 18S rRNA primers as an RNA 'loading control', given the difficulties of accurately measuring RNA concentration by methods such as spectrophotometry. In this way, the signal obtained from 18S rRNA is used as a measure of the total RNA in the sample. Once samples have been standardised in this way, there is greater confidence that differences in viral signal are actually due to differences in viral load and not due to differences in RNA loading or degradation.

References

1. Kawai A, Morimoto K (1994) Functional aspects of lyssavirus proteins. In: Rupprecht CE, Dietzschold B, Koprowski H (eds) Lyssaviruses. Springer, Berlin Heidelberg New York, pp 27–42
2. Wunner WH (1991) The chemical composition and structure of rabies viruses. In: Baer GM (ed) The natural history of rabies, 2nd edn. CRC Press, Boca Raton, LA, pp 31–67
3. Smith J, McElhinney LM, Heaton PR, Black EM, Lowings JP (2000) Assessment of template quality by the incorporation of an internal control into a RT-PCR for the detection of rabies and rabies-related viruses. J Virol Methods 84:107
4. Heaton PR, Johnstone P, McElhinney LM, Cowley R, O'Sullivan E, Whitby JE (1997) A hemi-nested PCR assay for the detection of six genotypes of rabies and rabies-related viruses. J Clin Microbiol 35:2762–2766

Development of One-Step RT-PCR for the Detection of an RNA Virus, Dengue, on the Capillary Real-Time Thermocycler

Boon-Huan Tan*, E'ein See, Elizabeth Lim, Eric Peng-Huat Yap

Introduction

Human infection with Dengue virus, an arthropod-borne virus, is a common and re-emerging disease in many tropical regions. In its most severe form, Dengue haemorrhagic fever, the resulting virus-induced coagulopathy is potentially lethal. However, classic clinical signs of Dengue infection appear only late, in the post-febrile phase of infection, and many subclinical episodes of infection remain asymptomatic and difficult to detect. There is a need to develop rapid diagnostic tests to detect early and subclinical Dengue infection for treatment on the individual level, and for preventive public health measures. Although the gold standard for Dengue detection depends on the serological detection of Dengue-specific antibodies and/or Dengue virus isolation by cell culture [1], molecular-based detection methods such as conventional polymerase chain reaction (PCR) assays have been published [2, 3, 4, 5, 6, 7]. While sensitive, these assays involve many hands-on steps and are hence costly, manual and prone to carryover contamination and errors.

By contrast, real-time PCR assays are quicker and do not require post-PCR manipulation. The progress of PCR and the detection of PCR products can be monitored together using real-time analysis. Real-time PCR with Taqman probes has been reported for Dengue viruses [8, 9]. Here, we present a one-step reverse transcription (RT)-PCR assay for the rapid detection of Dengue viruses using a single-labelled hybridization probe on the LightCycler instrument.

Materials

Equipment

LightCycler instrument (Roche Diagnostics, Mannheim, Germany)
LightCycler software vers.3.39 (Roche Diagnostics)
LightCycler Capillaries (Roche Diagnostics)
LightCycler Centrifuge Adapters (Roche Diagnostics)
LightCycler Cooling Block (Roche Diagnostics)
Primer Select software (DNAStar Inc.)

* Boon-Huan Tan (✉) (e-mail: nmiv2@nus.edu.sg)
Defence Medical Research Institute, Defence Science & Technology Agency,
Clinical Research Centre, MD11, #02-12, 10 Medical Drive, Singapore 117597

Reagents

Amplification Primers (GENSET Singapore Pte Ltd., Singapore)
Hybridisation probe (GENSET Singapore Pte Ltd.)
QIAamp RNeasy Mini Kit (Qiagen, Hilden, Germany)
Tth Polymerase (Roche Diagnostics)
Bovine Serum Albumin (Roche Diagnostics)
SYBR Green I (Molecular Probes, Eugene, USA)

Procedure

Materials

Dengue virus type 1 strain Hawaiian, Dengue virus type 3 strain H87, and Dengue virus type 4 strain H241 were obtained from Dr Jean Paul, Pasteur Institute, Cambodia. Dengue virus type 2 strain New Guinea C, Dengue virus type 2 clinical strains ST and 8771/93, yellow fever virus strain 17D, Japanese encephalitis virus strain Nakayama were obtained from the Department of Microbiology, National University of Singapore. Dengue virus type 2 clinical strains 4169/99 and 3386/99 were obtained from Singapore General Hospital. These virus strains were propagated in Vero E6 cells and plaque titration carried out using the standard method.

Sample Preparation

Virus RNA was extracted from the tissue culture fluid of Dengue-infected Vero cells with an activated membrane-based kit from Qiagen. The RNA was eluted in RNAse-free water and stored at –80°C. Purified viral RNAs from Dengue virus type 2 strain New Guinea C were used to optimise the RT-PCR assay conditions on the LightCycler. Once the assay was optimised, validation was achieved using four virus clinical strains (ST, 8771/93, 4169/99 and 3386/99) as indicated in Table 1. The specificity of the Dengue virus type 2 primers were tested against viral RNAs extracted from Dengue virus type 1 strain Hawaiian, Dengue virus type 3 strain H87, Dengue virus type 4 strain H241, Japanese encephalitis and yellow fever viruses (Table 1).

Table 1. List of viruses tested in the RT-PCR assay

Virus	Strains	Origin
Dengue virus type 1	Hawaiian	Hawaii
Dengue virus type 2	New Guinea C	New Guinea
Dengue virus type 2	ST	Singapore
Dengue virus type 2	8771/93	Singapore
Dengue virus type 2	4169/99	Singapore
Dengue virus type 2	3386/99	Singapore
Dengue virus type 3	H87	Philippines
Dengue virus type 4	H241	Philippines
Yellow fever	17D	Ghana
Japanese encephalitis	Nakayama	Japan

Oligonucleotides

PCR primers and hybridization probes were selected by scanning the sequence from Dengue virus type 2 strain New Guinea-C for a suitable target using the Primer Select software (DNAStar Inc.). The specificity of the selected sequences was confirmed by blasting against the GenBank database. In our RT-PCR assay, we use a single labelled sequence-specific probe for hybridization to RT-PCR products. The hybridization assay is a modification of the Bi-probe system, which uses Cy5-labelled probes in conjunction with SYBR Green I. Instead of labelling the sequence-specific probe at its 5′ end with Cy5, we labelled the Dengue virus type 2 specific probe at its 5′ end with LCRed640 (Table 2).

LightCycler RT-PCR

The following master mix was used with SYBR Green I and hybridization probe detection:

	Volume [μl]	[Final]
RT-PCR buffer (5×)	2	1×
dNTPs (3 mM each)	1	0.3 mM each
BSA (5 μg/μl)	1	0.5μg/μl
MnOAc (25 mM)	1	4 mM
Primers (9 μM each)	0.5+0.5	0.45μM each
Probe (4 μM)	0.5	0.2μM
SYBR Green I (20×)	0.5	1×
Tth polymerase (5 U/μl)	0.5	2.5 U
H_2O	1.5	

Table 2. Oligonucleotides

Dengue virus type 2 New Guinea C (GenBank Accession #M29095)				
	Position	Length	GC (%)	T_m (°C)
Primers (5′ to 3′)				
CCTAGACATAAT CGG G	8,388–8,403	16	50	48
GTGGTCTTGGTCATA G	8,463–8,449	16	41.2	48
Product		76 bp		
Probe				
LCRed640AGAAAAAATAAAACAAGAGC	8,411–8,431	20	30	50

The following RT-PCR protocol was used for both the SBYR Green I and hybridization probe detection:

- Reverse transcription for 15 min at 50°C
- Denaturation for 5 min at 95°C
- Amplification using 8-cycle PCR profile

Parameter	Value	
Cycle	8	
Type	Quantification	
	Segment 1	Segment 2
Target temperature [°C]	95	55
Incubation time [s]	0	7
Temperature transition rate [°C/s]	20	20
Acquisition mode	None	Single

- Amplification using 50-cycle PCR profile

Parameter	Value	
Cycle	50	
Type	Quantification	
	Segment 1	Segment 2
Target temperature [°C]	87	55
Incubation time [s]	0	7
Temperature transition rate [°C/s]	20	20
Acquisition mode	None	Single

- Melting Curve Analysis

Parameter	Value		
Cycles	1		
Type	Melting curve analysis		
	Segment 1	Segment 2	Segment 3
Target temperature [°C]	95	55	90
Incubation time [s]	0	30	0
Temperature transition rate [°C/s]	20	20	0.2
Acquisition mode	None	None	Continuous

Gel Electrophoretic Analysis of RT-PCR Products

The RT-PCR products were recovered from the capillaries by reverse centrifugation into microcentrifuge tubes. Samples were analysed on 2% agarose gels, stained with ethidium bromide and visualised under UV light.

Results

Viral RNA was extracted from Dengue virus type 2, New Guinea C, and used to optimise the RT-PCR assay on the LightCycler. Fig. 1 shows the RT-PCR results using tenfold dilutions of vRNA from 10 pfu (plaque-forming units) to 0.001 pfu per assay. Fig. 1A shows the quantitative analysis of RT-PCR products using SBYR Green I detection. The detection limit was consistently observed to be 0.01 pfu per PCR assay. The same detection limit was also observed in the presence of hybridization probe (Fig. 1B). For the RT-PCR assay containing 10 pfu, a significant increase in amplified products was observed from cycle 8 and a plateau was reached at cycle 20. In the RT-PCR assay containing 0.01 pfu, an increase in amplification products was only observed at cycle 18. Analysis of the melting curve profile of the RT-PCR

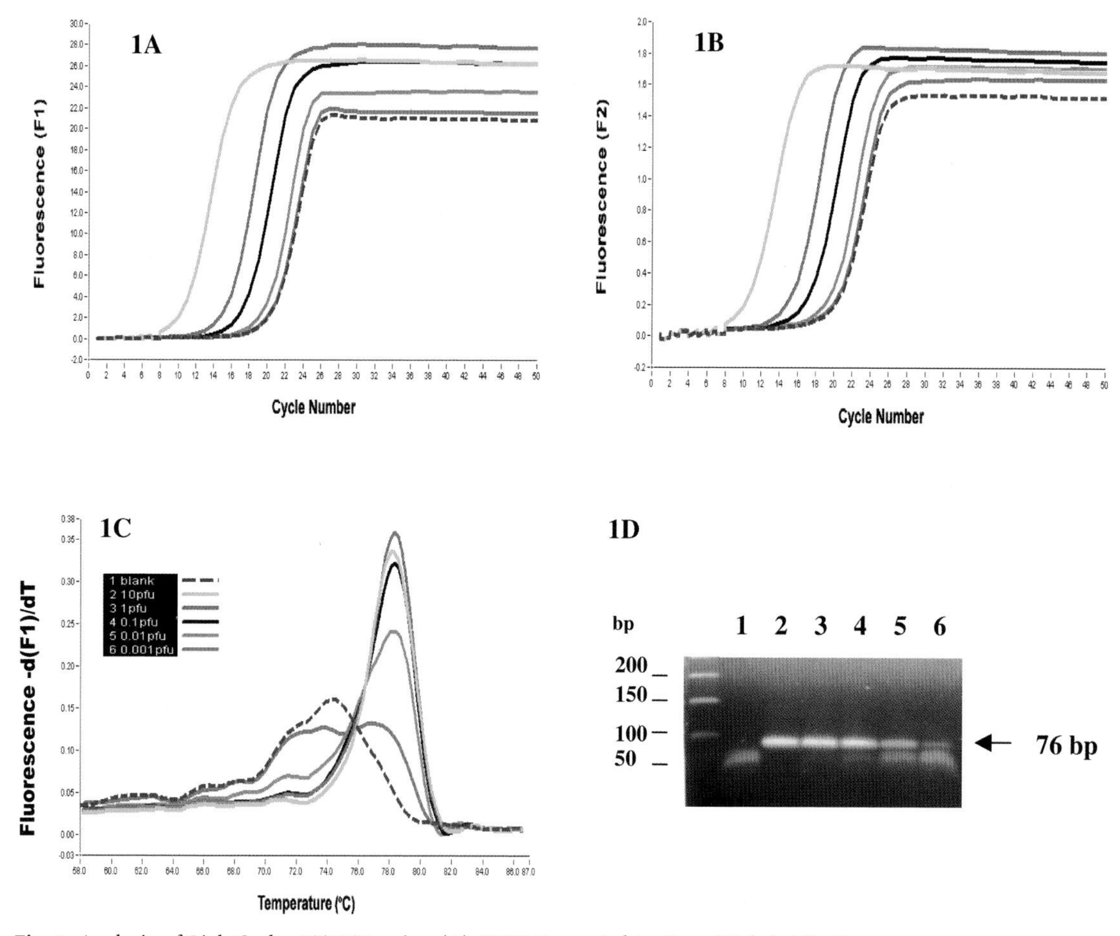

Fig. 1. Analysis of LightCycler RT-PCR using (**A**) SYBR Green I detection, (**B**) hybridization probe, (**C**) melting curve analysis, and (**D**) agarose gel electrophoresis

products using the negative derivative (−dF/dT) indicates the RT-PCR products peak at about 79°C (Fig. 1C). A second peak at 74°C in the samples containing water corresponds to the primer-dimer formation. A similar peak was also observed in the assay containing low copies of template RNA at 0.001 pfu. When the RT-PCR products were analysed on 2% agarose gel, as expected, an amplified product of size 76 bp was observed from the assay containing 10 pfu to 0.001 pfu (Fig. 1D, lanes 2–6, respectively). A band representing primer-dimers at around 50 bp was observed in lane 1 containing water instead of template RNA in the RT-PCR assay. The same band representing primer-dimers was also observed in lanes 4, 5 and 6, where low copies of RNA templates were present in the RT-PCR assay.

The RT-PCR assay was validated with viral RNA extracted from four clinical strains of Dengue virus type 2: ST, 8771/93, 4169/99 and 3386/99 (Table 1). These virus strains were isolated from Dengue-infected patients by conventional cell culture and serotyped with type-specific monoclonal antibodies by immunofluorescence assay. The RT-PCR assay was observed to be positive for all four strains of Dengue virus type 2 using SBYR Green I detection (Fig. 2A). The RT-PCR results were further confirmed with hybridization using a Dengue virus type 2-specific probe (Fig. 2B) and melting curve analysis (Fig. 2C). The RT-PCR products were further analysed on 2% agarose gel electrophoresis and amplified products, representing 76 bp, were observed in lanes 3–6, respectively, for all clinical samples: ST, 8771/93, 4169/99 and 3386/99 (Fig. 2D). No positive products were observed in lane 1 containing water in the RT-PCR assay.

The specificity of the RT-PCR assay was confirmed by replacing the Dengue virus type 2 RNA template with viral RNAs extracted from Dengue virus type 1, Dengue virus type 3, Dengue virus type 4, yellow fever and Japanese encephalitis viruses. No positive products were observed (results not shown).

Comments

Although detection with specific hybridization probes in the F2 channel served to confirm the specificity of the assay, our studies indicated that SYBR Green I detection in the F1 channel and melting curve analysis of RT-PCR products were able to confirm the sensitivity and specificity of the RT-PCR assay.

Primer-dimer formation was also observed in the assays containing no RNA template and in assays containing low copies of Dengue virus type 2 RNAs such as 0.001 pfu per assay. Primer-dimers were able to reduce the sensitivity of the RT-PCR assay and its signals sometimes gave false positive results. However, the melting curve analysis performed at the end of the RT-PCR assay differentiated the primer dimers, which dissociate at a lower temperature of 74°C, from the specific RT-PCR products, which dissociate at 79°C. It may also be possible to reduce primer-dimer formation by the use of a hot-start RT-PCR method.

The one-step RT-PCR assay for Dengue virus type 2 provides a rapid method for RNA detection. RT-PCR assays for Dengue virus types 1, 3 and 4 have also been developed (data not shown). The complete procedure of RNA detection including extraction of RNAs, reverse transcription, amplification and detection,

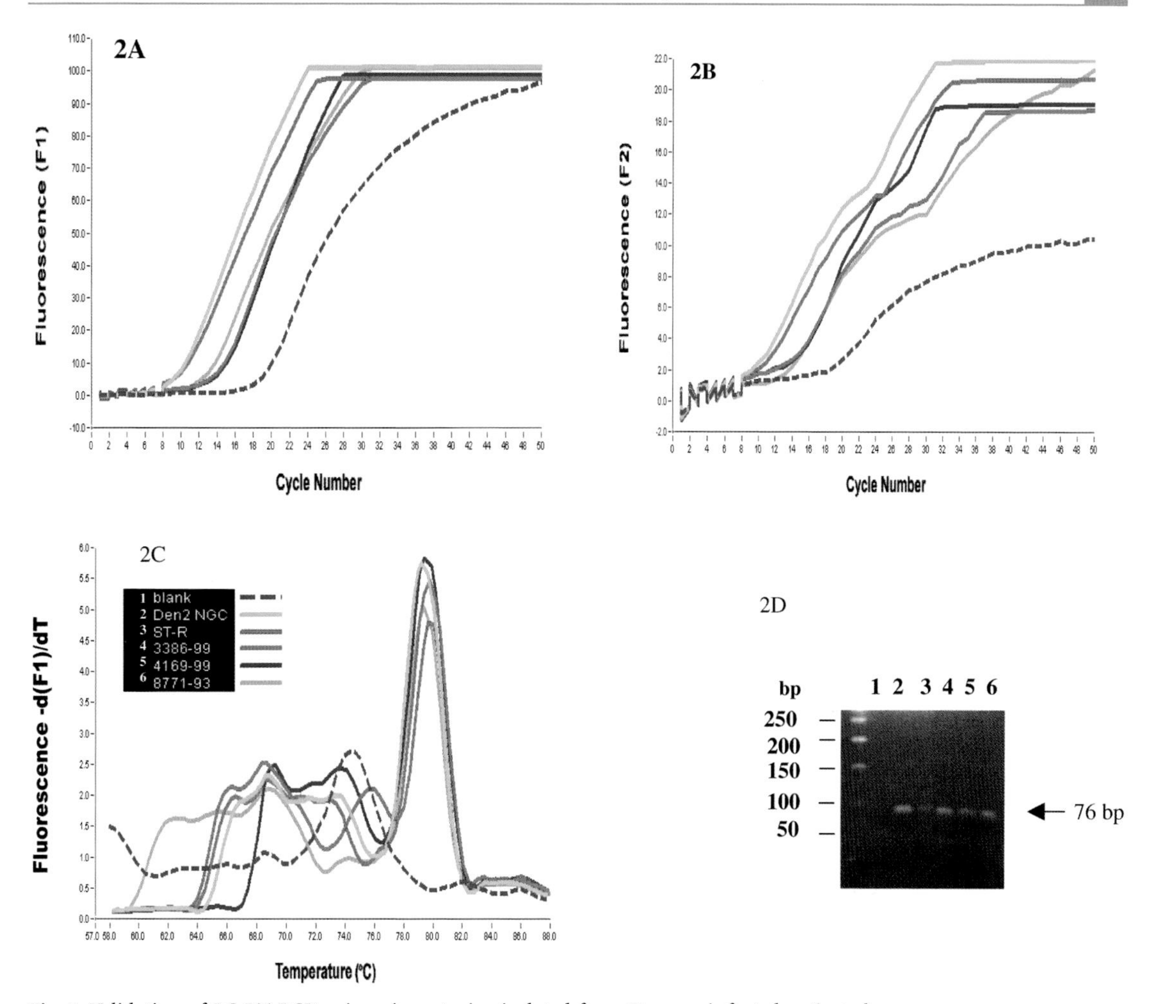

Fig. 2. Validation of LC RT-PCR using virus strains isolated from Dengue-infected patients by conventional cell culture using (**A**) SYBR Green I detection, (**B**) hybridization probe, (**C**) melting curve analysis, and (**D**) agarose gel electrophoresis

took about 1 h. The real-time PCR allowed direct monitoring of RT-PCR products as the assay progressed. No post-PCR manipulation was required. The direct fluorogenic detection of products from the sealed tube also avoided DNA contamination. In the routine diagnosis of clinical samples, the substitution of dUTP for dTTP combined with the use of Uracil-DNA glycosylase in the reaction mix could further reduce carryover contamination risk.

In conclusion, the one-step RT-PCR assay on the LightCycler instrument provides a highly sensitive, specific and rapid method to detect Dengue viral nucleic acids.

References

1. Gubler DJ (1989) Surveillance for dengue and dengue haemorrhagic fever. Bull Pan Am Health Organ 23:397–404
2. Henchal EA, Polo SL, Vorndam V, Yaemsiri C, Innis BL, Hoke CH (1991) Sensitivity and specificity of a universal primer set for the rapid diagnosis of dengue virus infections by polymerase chain reaction and nucleic acid hybridisation. Am J Trop Med Hyg 45:418–428
3. Tanaka M (1993) Rapid identification of flaviviruses using the polymerase chain reaction. J Virol Met 41:311–322
4. Pierre V, Drouet MT, Deubel V (1994) Identification of mosquito-borne flavivirus sequences using universal primers and reverse transcription/polymerase chain reaction. Res Virol 145:93–104
5. Meiyu F, Huosheng C, Cuihua C, Xiaodong T, Lianhua J, Yifei P, Weijun C, Huiyu C (1997) Detection of flaviviruses by reverse transcription polymerase chain reaction with the universal primer set. Microbiol Immunol 41:209–213
6. Sudiro TM, Ishiko H, Green S, Vaughn DW, Nisalak A, Kalayana-Rooj S, Rothman AL, Raengsakulrach B, Janus J, Kurane I, Ennis FA (1997) Rapid diagnosis of Dengue viremia by reverse transcriptase-polymerase chain reaction using 3′ noncoding region universal primers. Am J Trop Med Hyg 56:424–429
7. Harris E, Roberts TG, Smith L, Selle J, Kramar LD, Valle S, Sandoval E, Balmaseda A (1998) Typing of Dengue viruses in clinical specimens and mosquitoes by single-tube multiplex reverse transcriptase PCR. J Clin Microbiol 36:2634–2639
8. Laue T, Emmerich P, Schmitz H (1999) Detection of Dengue virus RNA in patients after primary or secondary dengue infection by using Taqman automated amplification system. J Clin Microbiol 37:2543–2547
9. Houng HH, Hritz D, Kanesa-Thasan N (2000) Quantitative detection of Dengue 2 virus using fluorogenic RT-PCR based on 3′ noncoding sequence. J Virol Method 86:1–11
10. Houng HH, Chen RC, Vaughan DW, Kanesa-Thasan N (2001) Development of a fluorogenic RT-PCR system for the quantitative identification of Dengue virus serotypes 1–4 using conserved and serotype-specific 3′ noncoding sequences. J Virol Method: 19–32

Detection of Genetically Modified Organisms (GMO) IV

Variation Analysis of Seven LightCycler Based Real-Time PCR Systems to Detect Genetically Modified Products (RRS, Bt176, Bt11, Mon810, T25, Lectin, Invertase) 251
Isabelle Dahinden, Andreas Zimmermann, Marianne Liniger, Urs Pauli

Variation Analysis of Seven LightCycler Based Real-Time PCR Systems to Detect Genetically Modified Products (RRS, Bt176, Bt11, Mon810, T25, Lectin, Invertase)

Isabelle Dahinden, Andreas Zimmermann,
Marianne Liniger, Urs Pauli*

Introduction

Foodstuffs, additives, and processing aids which comprise or are derived from more than 1% of genetically modified organisms (GMOs) have to be labeled according to Swiss [1] and EU law [2–4] as GMO products. As cultivation of GMOs is increasing rapidly and the products composed of GMO plants appear more and more on the world market, there is a need for sensitive, specific, and quick detection methods in order to monitor the labeling of foodstuffs. Several qualitative [5–7] as well as quantitative-competitive [8–10] PCR methods already exist. However, quantitative real-time PCR methods, which no longer require internal standards, competitors, time-consuming PCR reactions, and gel electrophoresis have recently appeared [11]. One such system is LightCycler real-time PCR technology. Real-time PCR systems allow the analysis of the samples while amplification is still in progress. Results can be obtained in less than 1 h.

This work describes the development and evaluation of seven new real-time PCR systems using the LightCycler real-time PCR technology. Five of them allow quantification of the GMO content of soybean (Roundup Ready soybean-RRSoybean), or corn (Bt176, Bt11, Mon810, and T25). The other two are amplification controls (to demonstrate the amplification capacity) of DNA from corn (invertase) and soybean (lectin). Intra- and inter-assay precision was studied.

Materials

Equipment

GeneQuant II RNA/DNA Calculator (Pharmacia, Uppsala, Sweden)
LightCycler instrument (Roche Diagnostics, Mannheim, Germany)
LightCycler Capillaries (Roche Diagnostics)
LightCycler software vers. 3.0 (Roche Diagnostics)
OLIGO Primer Analysis Software Version 5.0 for Windows (NBI, Plymouth MN, USA)

Reagents

Wizard DNA Extraction Kit (Promega, Madison, WI, USA)
Certified Standard for RRSoybean (Fluka, Buchs, Switzerland)

* Urs Pauli (✉) (e-mail: urs.pauli@bag.admin.ch)
Swiss Federal Office of Public Health, CH-3003 Bern, Switzerland

Certified Standard for Bt176 Corn (Fluka)
Bt11 corn (Novartis); Mon810 corn (Monsanto); T25 corn (Aventis)
Conventional corn and soybean (Fluka)
Hybridization Probes (TIB MOLBIOL, Berlin, Germany)
LightCycler-DNA Master SYBR Green I (Roche Diagnostics)
LightCycler-DNA Master Hybridization Probes (Roche Diagnostics)
Primers (Microsynth, Balgach, Switzerland)

Procedure

Sample Preparation

Samples of approximately 200 mg were incubated with 860 µl of TNE buffer [10 mM Tris-HCl, pH 8.0, 150 mM NaCl, 2 mM EDTA, and 1% (w/v) SDS], 100 µl of 5 M guanidine hydrochloride and 40 µl of 20 mg/ml proteinase K (Merck, Darmstadt, Germany) on a thermomixer (Eppendorf, Hamburg, Germany) at 58°C for at least 3 h. Co-extracted RNA was digested with 5 µl of RNase A [10 mg/ml] (Promega) at 58°C for 5 min [12, 13]. After centrifugation (10 min at 4°C and 14,000 rpm), 500 µl of the supernatant was combined with 1 ml Wizard resin (Promega, no. A7141) in a 1.5-ml Eppendorf tube, mixed by inversion and filled into a Wizard minicolumn (Promega, no. A 7211). Vacuum was applied to draw the solution through the minicolumn. The minicolumns were washed once with 2 ml 80% (v/v) isopropanol, centrifuged at 14,000 rpm for 1 min and dried by evaporation for 10–15 min at room temperature. Finally, the DNA was eluted in 50 µl of 10 mM Tris-HCl, pH 9.0. Single-stranded DNA concentrations were determined spectrophotometrically using a GeneQuant II RNA/DNA Calculator (Pharmacia). Each measurement was repeated twice and the values were averaged to calculate the concentration of the DNA.

10, 5, 2, 1, 0.5 and 0.1% DNA of Bt176, Bt11, Mon810 and T25 were diluted in conventional corn DNA. RRSoybean was diluted in conventional soybean DNA. 200 ng DNA of each of these samples were amplified.

Primer and Hybridization Probe Design

Primers and probes are described in Table 1. The length of PCR fragments was 125 bp for invertase (position of primer: 1415–1432 and 1517–1539; accession #: AF 171874), 114 bp for lectin (position of primer: 1217–1237 and 1310–1330; accession #: K 00821), 119 bp for RRSoybean, 144 bp for Bt176, 119 bp for Bt11, 145 bp for Mon810, and 138 bp for T25.

Oligonucleotides were designed with the help of the OLIGO software (NBI) in order to obtain comparable T_m values, GC contents, and to avoid stable secondary structures and stretches of palindromic sequences.

The oligonucleotides and probes were stored at –20°C until use.

Table 1. Oligonucleotides

Sequence (5′–3′)	Length	GC (%)	T_m (C°)
Invertase			
Primers			
CGTTCAACTGCAGCACCA	18	55.6	65.4
GCCCTTGAGCAGGTAGAAGTACA	23	52.2	67.9
Probes			
GTTCGGCCTTCTCGTGCTGGC-F	21	66.7	72.1
LCRed640-GACGACGACTTGTCCGAGCAGACC	24	62.5	72.3
Lectin			
Primers			
CCTCTACTCCACCCCCATCCA	21	61.9	69.0
CCATCTGCAAGCCTTTTTGTG	21	47.6	64.7
Probes			
TTGCCAGCTTCGCCGCTTC-F	19	63.2	70.3
LCRed640-TTCAACTTCACCTTCTATGCCCCTGAC	27	48.1	69.7
RRSoybean			
Primers			
ACCGTCTTCCCGTTACCTTG	20	55.0	66.1
GCCGGGCGTGTTGAG	15	73.3	65.4
Probes			
GCCGATGGCCTCCGCACA-F	18	72.2	71.9
LCRed640-GAAGTCCGCCGTGCTGCTCG	20	70.0	72.2
Bt176			
Primers			
GCGACTGGATCAGGTACAA	19	52.6	63.5
ACGGGGTTGGTGTAAATCT	19	47.4	63.1
Probes			
GGACATCGTGAGCCTGTTCCC-F	21	61.9	68.8
LCRed640-AACTACGACAGCCGCACCTACCC	23	60.9	72.0
Bt11			
Primers			
TATCATCGACTTCCATGACCA	21	42.9	62.6
GCAGATTACGCGCAGAAA	18	50.0	62.9
Probes			
CGTGAGTTTTCGTTCCACTGAG-F	22	50.0	65.2
LCRed640-GTCAGACCCCGTAGAAAAGATCAA	24	45.8	65.7
Mon810			
Primers			
ACTATCCTTCGCAAGACCCTTC	22	50.0	65.9
TTCAGAGAAACGTGGCAGTAACA	23	43.5	66.3
Probes			
GGAGAGGACACGCTGACAAGCTG-F	23	60.9	70.8
LCRed640-CTCTAGCAGATCTACCGTCTTCGGTACG	28	53.6	70.4
T25			
Primers			
GATTTCAGCGGCATGCCTGC	20	60.0	69.3
TGCGCGCAGCTGGATACAA	19	57.9	69.6
Probes			
CCTTGGAGGAGCTGGCAACTCAAAATCC-F	28	53.6	72.6
LCRed640-TTTGCCAAAAACCAACATCATGCCATCC	28	42.9	70.5

LightCycler PCR

A mastermix with the following reaction components was used for the amplification:

	Volume [μl]	[Final]
LightCycler-DNA Master Hybridization Probes	2	1X
$MgCl_2$ (25 mM)	2.4	4 mM
Primers (20 μM each)	0.5+0.5	0.5 μM each
Probes (10 μM each)	0.8+0.8	0.4 μM each
H_2O (PCR grade)	3 μl	
Total volume	10 μl	

To complete the amplification mixtures, 10 μl mastermix and 10 μl (200 ng) template DNA of the corresponding dilutions, negative (H_2O) as well as positive controls (genomic DNA) were added to each capillary. After a short centrifugation, the glass capillaries were placed into the LightCycler rotor.

- Denaturation at 95°C for 30 s (conventional PCR) or 10 min (FastStart PCR)
- Amplification

Parameter				Value		
Cycles				60		
Type				Quantification	Acquisition mode	Gains
	Fast Start	Target temperature [°C] Segments 1; 2; 3	Incubation time (s) Segments 1; 2; 3	Temperature transition rate [°C/s] Segments 1; 2; 3		
Invertase	Yes	95; 55; 72	10; 10; 6	20; 20; 2	n; s; n	F1=1; F2=25; F3=30
Lectin	No	95; 62; 72	0; 8; 6	20; 20; 20	n; s; n	F1=10; F2=25; F3=30
Bt176	No	95; 55; 72	0; 7; 6	20; 20; 2	n; s; n	F1=10; F2=30; F3=30
Bt11	Yes	95; 58; 72	0; 8; 6	20; 20; 20	n; s; n	F1=10; F2=27; F3=30
Mon810	Yes	95; 60; 72	0; 8; 6	20; 20; 2	n; s; n	F1=10; F2=26; F3=30
T25	Yes	95; 60; 72	0; 8; 6	20; 20; 20	n; s; n	F1=10; F2=26; F3=30
RRS	Yes	95; 63; 72	0; 8; 6	20; 20; 2	n; s; n	F1=10; F2=28; F3=30

n, none; s, single

Results

First, functionality of primers was tested in advance with SYBR Green I to exclude primer dimer formation. Then, concentrations of $MgCl_2$, primers and probes, incubation temperatures and times were optimized and the specificities for the different systems were tested. Except for the lectin and the Bt176 systems, a FastStart PCR protocol was used because of higher sensitivity. The sensitivity was less than 0.1% of the desired target DNA in all seven systems.

The corn-specific invertase gene and the soybean-specific lectin gene served as a control for the amplification capacity of the isolated DNA by PCR. A calibration curve was drawn using standard concentrations of 10, 5, 2, 1, 0.5, and 0.1% GMO DNA (RRSoybean, Bt176, Bt11, Mon810, and T25 corn) which were diluted in conventional soybean or corn DNA, respectively. Precision was assessed by intra- and inter-assay studies. Six different concentrations of each target were analyzed in four replicates in the same LightCycler run. On 5 subsequent days, the DNA extraction and LightCycler study was repeated. One dilution series was used to draw the calibration curve by plotting the crossing point against the log of the genome copies. Then, the coefficient of variation (100 x mean/standard deviation) at each concentration was calculated. Table 2 lists the average intra-assay coefficients of variation for each target at each concentration (including minimal and maximal obtained values = range).

In general, high variation (>20%) did not correlate with the GMO content. Very high variation values (>30%) were due to one or two results out of the range of the standard curve. As expected, inter-assay variation is higher than intra-assay variation, possibly because of differences between different extractions, solutions of primers and probes, or PCR amplification.

Comments

The present study describes seven new LightCycler systems for quantifying the amount of corn (invertase system), soybean (lectin system), Bt176, Bt11, Mon810, T25, and RRSoybean in foodstuffs. These systems were developed in order to monitor genetically modified organisms in food. The CV was between 5%–15% in most assays. The sensitivity of these assays was at least 0.1% genetically modified DNA diluted in conventional DNA.

Table 2. Variation analysis of intra-assays for lectin, invertase, RRSoybean, Bt176, Bt11, Mon810, and T25. Variation (%): overall variation of all assays. Range (min-max %): minimal and maximal variation values in per cent for every set.

	GMO (%)	variation (%)	range (min-max %)
Lectin	10	17.7	9.4–25.3
	5	11.5	2.8–13.1
	2	14.9	6.4–23.0
	1	14.9	5.2–21.6
	0.5	19.4	11.6–23.4
	0.1	21.9	9.7–29.4
Invertase	10	9.5	4.9–9.4
	5	10.9	4.7–13.2
	2	13.4	4.5–15.1
	1	9.8	3.4–7.9
	0.5	12.7	5.1–15.0
	0.1	17.3	6.1–21.1
RRS	10	23.8	13.5–27.9
	5	21.7	14.1–21.4
	2	19.3	8.9–22.6
	1	19.0	7.4–19.4
	0.5	30.9	16.0–30.5
	0.1	43.5	19.9–37.0
Bt176	10	14.4	5.7–18.3
	5	16.0	9.7–17.8
	2	14.5	3.9–11.5
	1	14.2	9.8–16.5
	0.5	16.6	5.5–20.3
	0.1	16.5	4.8–18.2
Bt11	10	18.1	5.4–24.2
	5	30.3	6.2–40.6
	2	13.4	6.2–15.6
	1	13.1	6.6–16.0
	0.5	18.9	3.0–22.2
	0.1	15.5	4.3–23.0
Mon810	10	19.2	4.1–22.1
	5	17.3	5.6–13.8
	2	12.0	8.0–13.3
	1	28.3	3.5–20.1
	0.5	19.5	7.6–26.2
	0.1	33.4	24.6–31.5
T25	10	12.0	6.1–11.7
	5	11.3	6.9–12.8
	2	12.2	5.1–16.6
	1	13.4	2.8–24.4
	0.5	27.9	7.2–58.9
	0.1	28.3	17.0–23.3

Table 3. Variation analysis of inter-assays for lectin, invertase, RRSoybean, Bt176, Bt11, Mon810, and T25. Variation (%): overall variation of all assays. Range (min-max %): minmal and maximal variation values in per cent for every set.

	GMO (%)	variation (%)	range (min-max %)
Lectin	10	19.8	7.7–26.1
	5	12.8	8.8–14.4
	2	16.7	7.1–17.5
	1	16.5	9.3–18.0
	0.5	23.3	10.5–22.2
	0.1	21.9	10.1–21.3
Invertase	10	10.5	4.6–15.1
	5	12.2	5.5–14.8
	2	15.0	5.6–16.3
	1	11.0	7.4–11.4
	0.5	15.3	3.6–20.8
	0.1	17.3	2.7–24.6
RRS	10	25.7	14.8–24.4
	5	24.5	6.1–21.1
	2	21.5	8.7–25.8
	1	21.7	9.6–26.3
	0.5	37.1	8.2–32.8
	0.1	43.5	17.0–63.4
Bt176	10	15.8	7.1–21.9
	5	18.3	7.6–18.9
	2	16.0	9.7–14.9
	1	16.1	8.5–13.2
	0.5	19.8	8.5–16.6
	0.1	16.5	5.6–27.9
Bt11	10	19.7	4.3–22.6
	5	34.8	9.0–37.9
	2	14.8	6.7–16.3
	1	14.5	4.1–20.4
	0.5	22.5	6.1–25.5
	0.1	15.5	2.3–20.6
Mon810	10	21.3	5.3–28.7
	5	19.8	8.2–19.7
	2	13.2	1.5–16.6
	1	31.8	9.2–31.4
	0.5	23.3	12.6–21.4
	0.1	33.4	10.2–42.2
T25	10	13.2	7.9–13.5
	5	12.8	4.4–16.0
	2	13.5	7.1–14.9
	1	15.2	10.9–15.0
	0.5	33.1	5.7–49.3
	0.1	28.3	9.7–23.8

References

1. Lebensmittelverordnung (Swiss Food Ordinance) of 1 March 1995, SR 817.02. Eidgenössische Drucksachen und Materialzentrale (EDMZ), 3003 Bern, Switzerland
2. Commission Regulation (EC) no. 258/97 of the European Parliament and the Council of 26 January 1997 concerning Novel Foods and Novel Food Ingredients. Official Journal of the European Communities L43:1–5
3. Commission Regulation (EC) no. 49/2000 of 11 January 2000. Official Journal of the European Communities L6:13–14
4. Commission Regulation (EC) no. 50/2000 of 10 January 2000. Official Journal of the European Communities L6:15–16
5. Studer E, Dahinden I, Lüthy J, Hübner P (1997) Nachweis des gentechnisch veränderten "Maximizer"-Mais mittels der Polymerase-Kettenreaktion (PCR). Mitt Gebiete Lebensm Hyg 88:515–524
6. Van Hoef AMA, Kok EJ, Bouw E, Kuiper HA, Keijer J (1998) Development and application of a selective detection method for genetically modified soy and soy-derived products. Food Addit Contam 15:767–774
7. Zimmermann A, Hemmer W, Liniger M, Lüthy J, Pauli U (1998) A sensitive detection method for genetically modified MaisGard corn using a nested PCR-system. Z Lebensm Unters Forsch 31:664–667
8. Studer E, Ryhner C, Lüthy J, Hübner P (1998) Quantitative competitive PCR for the detection of genetically modified soybean and maize. Z Lebensm Unters Forsch 207:207–213
9. Hardegger M, Brodmann P, Herrmann A (1999) Quantitative detection of the 35S promotor and the NOS terminator using quantitative competitive PCR. Z Lebensm Unters Forsch 209:83–87
10. Zimmermann A, Lüthy J, Pauli U (2000) Event specific transgene detection in Bt11 corn by quantitative PCR at the integration site. Z Lebensm Wiss Technol 33:210–216
11. Dahinden I, Stadler M, Pauli U, Lüthy J (2000) Evaluation of a real-time PCR to detect coeliac toxic components and comparison to the ELISA method analysing 35 baby food samples. Mitt Gebiete Lebensm Hyg 91:723–732
12. Zimmermann A, Lüthy J, Pauli U (1998) Quantitative and qualitative evaluation of nine different extraction methods for nucleic acids on soybean food samples. Z Lebensm Unters Forsch A 207:81–90
13. Swiss Food Manual (2001) Immunchemische und molekularbiologische Methoden: Isolation und Reinigung von Nucleinsäuren aus Lebensmitteln mittels Wizard. Chapter 52B. EDMZ, 3003 Bern, Switzerland (CD-ROM 311.510)